HELICOBACTER PYLORI - NEW APPROACHES OF AN OLD HUMAN MICROORGANISM

Edited by **Bruna Maria Roesler**

Helicobacter Pylori - **New Approaches of an Old Human Microorganism**
http://dx.doi.org/10.5772/intechopen.73711
Edited by Bruna Maria Roesler

Contributors

Tadeusz Lapinski, Amirhossein Sheikhshahrokh, Elnaz Saeidi, Leif Percival Andersen, Rie Louise Møller Nordestgaard, Malene Roed Spiegelhauer, Agnes Tving Stauning, Tove Havnhøj Frandsen, Ida Caroline Gren, Carolina Romo Gonzalez, Rafael Coria-Jimenez, Felicia Galos, Cătălin Boboc, Anca Orzan, Cristina Coldea, Mălina Anghel, Mihaela Bălgrădean, M. Cristina L. Martins, Paula Parreira, Catarina Seabra, Daniela Lopes-De-Campos, Bruna Maria Roesler

First published in London, United Kingdom, 2019 by IntechOpen
IntechOpen is the global imprint of INTECHOPEN LIMITED, registered in England and Wales, registration number: 11086078, The Shard, 25th floor, 32 London Bridge Street
London, SE19SG – United Kingdom
Printed in Croatia

British Library Cataloguing-in-Publication Data
A catalogue record for this book is available from the British Library

Additional hard and PDF copies can be obtained from orders@intechopen.com

Helicobacter Pylori - New Approaches of an Old Human Microorganism, Edited by Bruna Maria Roesler
p. cm.
Print ISBN 978-1-83881-146-4
Online ISBN 978-1-83881-147-1
eBook (PDF) ISBN 978-1-83881-148-8

For EU product safety concerns:
IN TECH d.o.o., Prolaz Marije Krucifikse Kozulić 3, 51000 Rijeka, Croatia,
info@intechopen.com or visit our website at intechopen.com.

Meet the editor

Dr. Bruna Maria Roesler is a pharmacist biochemist and holds a Master's Degree in Pharmacology and a Doctoral Degree in Basic Sciences – Internal Medicine from State University of Campinas (Campinas, SP, Brazil) where she has identified the principal genotypes of *Helicobacter pylori* strains in patients with chronic gastritis, peptic ulcer disease and gastric cancer (early and advanced stages) through molecular biology techniques. She has published her work in several peer-reviewed journals and given oral and poster presentations at various congresses. Her research also includes the etiology, the epidemiology, and the physiopathology of gastrointestinal diseases. She has also participated in some studies which reported the possible relationship between *H. pylori* and idiopathic thrombocytopenic purpura, as well as between *H. pylori* and liver diseases.

Contents

Preface

Helicobacter pylori (*H. pylori*) remains one of the most common worldwide human infections and, although its colonization is not a disease in itself, it is a condition that affects the relative risk of developing various clinical disorders of the upper gastrointestinal tract, such as chronic gastritis, peptic ulcer disease, mucosa-associated lymphoid tissue (MALT) lymphoma and gastric adenocarcinoma. Besides, in the last decades, the infection and its consequences produced by this microorganism has been associated with the development of various extradigestive disorders, such as hepatobiliary, cardiovascular and pancreatic diseases, iron deficiency anemia, idiopathic thrombocytopenic purpura, skin diseases, among others.

Despite the bacteria infects half of the world population, most patients are asymptomatic for life, while only some will come to develop a digestive disease. Nevertheless, gastric cancer remains one of the principal causes of cancer death worldwide and *H. pylori* is consider as a class I carcinogen (International Agency for Research on Cancer – World Health Organization).

Transmission of *H. pylori* is still not entirely clarified, but human-to-human spread through oral-oral or fecal-oral route is thought to be most plausible. The infection is typically acquired during childhood and usually becomes a lifelong infection, if left untreated. The host certainly mounts an immune response, but it fails to clear the infection, allowing *H. pylori* to establish a persistent infection and a chronic inflammation.

In this book, compounded by eight chapters, important aspects of *H. pylori* were reported and the book divided into three following sections: "General aspects of *Helicobacter pylori* infection", "Virulence factors of *Helicobacter pylori*", and "*Helicobacter pylori* and eradication therapy".

The first section comprehends four chapters, "*Helicobacter pylori*: an overview of an old human microorganism", "Clinical manifestations of the *Epsilonproteobacter, Helicobacter pylori*", "Endoscopical aspects of *Helicobacter pylori* gastritis in children", and "The importance of *H. pylori* infection in liver diseases".

In the second section some important aspects regarding virulence factors of *H. pylori* were reported in two chapters, "VacA genotype" and "*Helicobacter pylori* genes jhp0940, jhp0945, jhp0947 and jhp0949 associated to gastroduodenal disease".

Finally, in the third section, two chapters explore aspects concerning the eradication treatment of *H. pylori* infection and possible resistance mechanisms to the antibiotics commonly used for this purpose, besides the use of natural medicines for eradication of this microorganism: "Gastric microbiota and resistance to antibiotics" and "Non-antibiotic based therapeutics targeting *Helicobacter pylori*: from Nature to the lab".

"*Helicobacter pylori* – New Approaches of an Old Human Microorganism" is a book which will certainly provide an updated set of information in important aspects of this microorganism that has co-evolved with humans for over 60,000 years.

The editor expresses her thankfulness for the excellent work of the contributing authors. The editor thanks the entire In Tech Open Access publishing team for all its attention and support, making possible the accomplishment of this book. The editor is especially thankful for the excellent support given by Ms. Lada Bozic in all the steps of this book.

Dr. Bruna Maria Roesler
Pharmacist Biochemist
Center of Diagnosis of Digestive Diseases
School of Medical Sciences, State University of Campinas
Campinas, São Paulo, Brazil

Section 1

General Aspects of *Helicobacter Pylori* Infection

Chapter 1

Introductory Chapter: *Helicobacter pylori* - An Overview of an Old Human Microorganism

Bruna Maria Roesler

Additional information is available at the end of the chapter

http://dx.doi.org/10.5772/intechopen.88806

1. Introduction

The human stomach is an unfriendly place for most infective bacteria probably due to the very low pH found in this place. However, the first isolation of a spiral-shaped, Gram-negative and microaerophilic bacterium in 1982 by Warren and Marshall [1] significantly changed the concepts of gastric microbiology.

Initially, this bacterium was named *Campylobacter pyloridis*, but analysis of nucleic acid sequence and ultrastructural studies besides the helical shape allowed differentiation of this genus to *Helicobacter*. Finally, the species was named *pylori* because it can be found most often in the antral mucosa, near the pylorus [2].

Helicobacter pylori (*H. pylori*) organisms are 2.5–5.0 μm long and 0.5–1.0 μm wide, with two to six unipolar-sheathed flagella, which are essential for bacterial motility [3]. It has been described that bacteria can exist in three different morphologic forms: the viable and culturable spiral form, the viable but nonculturable (VBNC) coccoid form which are less virulent, and the nonviable degenerative *H. pylori* form [4].

Colonization with *H. pylori* is commonly acquired during childhood and induces chronic gastritis in all infected individuals unless specific treatment is given [5, 6]. While over 80% of infected subjects remain asymptomatic [7], *H. pylori* chronic infection has been associated with the development of various clinical disorders of the upper gastrointestinal tract, such as chronic gastritis, peptic ulcer disease, mucosa-associated lymphoid tissue (MALT) lymphoma, and gastric adenocarcinoma [8]. In fact, *H. pylori* infection is a significant risk factor for the development of gastric cancer, and bacterium is classified as a group I carcinogen by the World Health Organization [9].

Although *H. pylori* is primarily responsible for the upper gastrointestinal diseases, only 10% of people colonized with this bacterium portray disease symptoms. It suggests that host and bacterial factors also contributed to differences in *H. pylori* pathogenicity [10, 11]. For instance, the risk of developing gastric cancer is also related to genetic characteristics of the host and environmental factors, which, associated with specific bacterial strain characteristics, influence the severity of the chronic inflammatory response [12, 13].

H. pylori is perhaps the most ubiquitous and successful human pathogen, since it colonizes the stomach of more than 50% of the world population [14, 15]. It has been demonstrated that *H. pylori* has a long period of coevolution with humans, going back at least since human migration out of Africa about 60,000 years ago [16, 17]. There are very well-characterized mechanisms of adaptation which was developed by ancestral *H. pylori* over the time. Through selection and coevolution, this bacterium established measures which actively and passively avoid the human immune response [18].

H. pylori infection results in recruitment of neutrophils, lymphocytes, and macrophages into the gastric mucosa through the induction of several cytokines such as TNF-α, IL-6, and IL-8 [19, 20]. It is believed that the immune response during infection plays an important role in the pathogenesis. *H. pylori* successfully establishes a chronic infection by achieving a delicate balance between inducing immune response and surviving in the inflammatory milieu by using an array of important virulence factors [15].

H. pylori presents important virulence factors which are essential both for bacterium colonization and maintenance in the human stomach (such as urease and flagella) and for the interaction with the gastric epithelial cells, the bacterial adhesins (blood group antigen-binding adhesion (BabA), sialic acid-binding adhesion (SabA), AlpA and AlpB, HopZ, and OipA). Besides, virulence factors involved in gastric inflammation are important for the development of chronic infection and clinical symptoms of gastrointestinal diseases (the principal are cytotoxin-associated gene-pathogenicity island (cagPAI), vacuolating cytotoxin A (VacA), and duodenal ulcer promoting gene (dupA)).

2. Epidemiology of *H. pylori* infection

The *H. pylori* infection has emerged as one of the most common chronic bacterial infections worldwide and affects more than half of the world's population, with clinical signs of infection only manifesting in <20% of these individuals [21].

H. pylori is thought to be indigenous to the human population and is well adapted to existing in the human stomach for the lifetime of its host [22] unless eradication using appropriate chemotherapeutic agents is successful. Lifelong colonization seems to be due to the ability of some strains of *H. pylori* to both adapt to the host's immunological responses and to also withstand the constantly changing gastric environment [23].

The rate of *H. pylori* infection differs among groups as well as within the population. Strains from different geographical areas exhibit phylogeographic features [24–26]. The genomic patters of *H. pylori* have been shown to be extremely diverse, and gastric mucosa may be colonized by strains with small differences in the genomic patterns suggesting subtype variation [27].

The prevalence of *H. pylori* infection varies widely by geographic area, age, race, and socioeconomic status. While the infection is on a fast decline in the most of the Western countries, mainly due to the success of therapeutic regimens and improved personal and community hygiene that prevents reinfection, in developing countries, the prevalence rates can reach 90% and is higher among individuals belonging to low socioeconomic status group [28, 29]. It occurs especially due to failure of treatment and emergence of drug resistance [25, 30].

Most studies suggest that males and females are infected at approximately the same rates [31–33]. In spite of it, a meta-analysis population-based study reported a male predominance of *H. pylori*-related diseases in adults but not in children [34].

The infection probably occurs in the childhood, and children are often infected by a strain with a genetic fingerprint identical to that of their parents [35]. Besides, local prevalence of *H. pylori* within a country also should be considered, and there are estimates that infection is more common in rural developing areas than in urban developed ones [36].

Moreover, differences by ethnic and racial groups are evident [31, 32, 37]. In addition, the main risk factors of *H. pylori* infection, especially if present during childhood, have been associated with socioeconomic status. Malaty and Graham [38] demonstrated that there is probably an inverse correlation between prevalence and socioeconomic status. It has also been reported that overcrowding, such as living in a crowded environment, sibship size, number of persons or children in the home, number of persons per room, crowding index, and living in an institutionalized population, is a situation consistently related to *H. pylori* positivity [39–42].

Finally, it is important to consider that the pathogenetic role of *H. pylori* in gastroduodenal pathologies has been elucidated and confirmed in the past 30 years [43] redirecting the scientific and medical understanding of great part of gastrointestinal diseases. The development of effective therapies against *H. pylori* infection has progressed, and its successful eradication leads to healing of chronic active gastritis and reverses inflammation of the mucosa. In spite of it, the challenge nowadays is gastric cancer and the understanding of gastric carcinogenesis, almost always associated with *H. pylori* long-term infection [44].

3. Transmission pathways

Although the natural niche for *H. pylori* is the human stomach, some questions about other possible reservoirs for bacterium have been appearing in the last years. Nevertheless, most part of the questions about the transmission of *H. pylori* remains unclear, and, because of it, the possible modes of transmission are still unknown. Consequently, the routes of transmission of *H. pylori* are supposed to occur via an array of different pathways.

Some important studies have reported and highlighted the importance of *H. pylori* biofilms, the presence of coccoid forms within the biofilm, and resistance, providing insight into the prevalence of coccoid forms in the gastric mucosa. These reports are very important because these can bring a better understanding about the mechanisms behind recalcitrant coccoid states and how they can phenotypically shift into more virulent spiral forms [21, 45–47].

The infection is typically acquired in early childhood and once established commonly persists throughout life unless treated. Person-to-person transmission within the family appears to be the predominant mode of transmission, particularly from mothers to children and among siblings, indicating that intimate contact is important [29, 48–50]. The route of transmission is uncertain, but the gastro-oral, oral-oral, and fecal-oral routes are likely possibilities.

The community and environment may play additional roles for *H. pylori* transmission in some settings. Molecular analyses show that the microorganism is also present in various aquatic environments suggesting that human-fecal-contaminated water sources could be a plausible reservoir of the pathogen. The persistence of the environment virulent *H. pylori* strain in a clustered state, such as the biofilm, suggests a long-term survival of the bacterial community outside the host, enabling bacterial transmission with important clinical repercussions [21, 46]. In addition, zoonotic transmission by houseflies [51–53] and some domestic animals such as dogs, cats, and sheep [54–56], as well as iatrogenic transmission [57, 58], have been proposed. Besides, there can be factors both from host and bacterium which may modify the acquisition and persistence of *H. pylori* infection.

Another possibility of *H. pylori* transmission which has been extensively reported is the water. The contamination of drinking water by human feces has been suggested as one of the possible routes of *H. pylori* transmission, and it has been demonstrated that the microorganism is present in the so-called viable but nonculturable state in this unsuitable environment, meaning that their role in fecal-oral transmission via contaminated water sources cannot be disregarded [47, 59]. The first evidences of water transmission route were obtained in studies developed in some Latin American countries—Peru, Colombia, Chile, and Venezuela—and since then *H. pylori* has been detected in several water sources, including lakes, rivers, tap water, well water, irrigation water, and sea water, and also in water distribution systems. Consequently, it can be hypothesized that drinking water could be the pathway for returning to humans [14]. Consequently, it can be suggested that water can serve as an intermediate source in the fecal-oral transmission of *H. pylori*, acting as a reservoir in which this pathogen can survive for long periods.

4. *H. pylori* eradication therapies

The principal cases in which *H. pylori* have to be eradicated have been discussed in several guidelines worldwide, also considering that this microorganism is sensitive to only a few medications, and their widespread use in other kind of infections has led to a reduction in their effectiveness against the bacterium.

The infection is typically treated with combinations of two to three antibiotics along with a proton pump inhibitor (PPI), taken concomitantly or sequentially for periods ranging from 3 to 14 days. In spite of it, there is no treatment regimen which guarantees cure of *H. pylori* infection in 100% of patients. Individuals should be asked about any previous antibiotic uses, information that has to be taken into consideration when choosing an *H. pylori* treatment regimen.

Clarithromycin triple therapy consisting of a PPI, clarithromycin, and amoxicillin or metronidazole for 14 days remains a recommended treatment option in regions where *H. pylori* clarithromycin resistance is known to be <15% and in patients with no previous history of macrolide exposure for any reason. Bismuth quadruple therapy consisting of a PPI, bismuth, tetracycline, and a nitroimidazole for 10–14 days is a recommended first-line treatment option. Concomitant therapy consisting of a PPI, clarithromycin, amoxicillin, and nitroimidazole for 10–14 days is a recommended first-line treatment option. Levofloxacin triple therapy consisting of a PPI, levofloxacin, and amoxicillin for 10–14 days is a suggested first-line treatment option. Finally, fluoroquinolone sequential therapy consisting of a PPI and amoxicillin for 5–7 days followed by a PPI, fluoroquinolone, and nitroimidazole for 5–7 days is a suggested first-line treatment option [60–62].

This book comprehends important chapters that will certainly clarify the understanding of this microorganism infection, which affects half of the world population, despite promoting clinical symptoms and disease in only a small part of the infected individuals.

Author details

Bruna Maria Roesler

Address all correspondence to: roeslerbruna@gmail.com

Center of Diagnosis of Digestive Diseases, School of Medical Sciences, State University of Campinas, Campinas, São Paulo, Brazil

References

[1] Marshall BJ, Warren JR. Unidentified curved bacilli in the stomach of patients with gastritis and peptic ulceration. Lancet. 1984;**1**:1311-1315

[2] Goodwin CS et al. Transfer of *Campylobacter pylori* and *Campylobacter mustelae* to *Helicobacter* gen. nov. as *Helicobacter pylori* comb. nov. and *Helicobacter mustelae* comb. nov., respectively. International Journal of Systematic Bacteriology. 1989;**39**:397-405

[3] Geis G, Leying H, Suerbaum S, et al. Ultrastructure and chemical analysis of *Campylobacter pylori* flagella. Journal of Clinical Microbiology. 1989;**27**:436-441

[4] Andersen LP, Rasmussen L. *Helicobacter pylori*-coccoid forms and biofilm formation. FEMS Immunology and Medical Microbiology. 2009;**56**:112-115

[5] Buck GE, Gourley WK, Lee WK, et al. Relation of *Campylobacter pyloridis* to gastritis and peptic ulcer. The Journal of Infectious Diseases. 1986;**153**:664-669

[6] Testerman TL, Morris J. Beyond the stomach: An updated view of *Helicobacter pylori* pathogenesis, diagnosis and treatment. World Journal of Gastroenterology. 2014;**20**: 12781-12808

[7] Blaser MJ. Who are we? Indigenous microbes and the ecology oh human diseases. EMBO Reports. 2006;**7**:956-960

[8] Kusters JG, van Vliet AHM, Kuipers EJ. Pathogenesis of *Helicobacter pylori* infection. Clinical Microbiology Reviews. 2006;**19**:449-490

[9] IARC Working Group on the evaluation of carcinogenic risks to humans. Schistosomes, liver flukes and *Helicobacter pylori*. Lyon, 7-14 June 1994. IARC Monographs on the Evaluation of Carcinogenic Risks to Humans. 1994;**61**:1-241

[10] Mégraud F. Impact of *Helicobacter pylori* virulence on the outcome of gastroduodenal diseases: Lessons from the microbiologist. Digestive Diseases. 2001;**19**:99-103

[11] Ahmed N, Sechi LA. *Helicobacter pylori* and gastroduodenal pathology: New threats of the old friend. Annals of Clinical Microbiology and Antimicrobials. 2005;**4**:1-10

[12] Peek RM, Blaser MJ, Mays DJ, et al. *Helicobacter pylori* strain specific genotypes and modulation of the gastric epithelial cell cycle. Cancer Research. 1999;**59**:6124-6131

[13] de Vries AC, Haringsma J, Kuipers EJ. The detection, surveillance and treatment of premalignant gastric lesions related to *Helicobacter pylori* infection. Helicobacter. 2007;**12**:1-15

[14] García A, Salas-Jara MJ, Herrera C, et al. Biofilm and *Helicobacter pylori*: From environment to human host. World Journal of Gastroenterology. 2014;**20**:5632-5638

[15] Lina TT, Alzabrani S, Gonzalez J, et al. Immune evasion strategies used by *Helicobacter pylori*. World Journal of Gastroenterology. 2014;**20**:12753-12766

[16] Falush D, Wirth T, Linz B, et al. Traces of human migration in *Helicobacter pylori* populations. Science. 2003;**299**:1582-1585

[17] Moodley Y, Linz B, Yamaoka Y, et al. The peopling of the Pacific from a bacterial perspective. Science. 2003;**323**:527-530

[18] Kalali B, Mejias-Luque R, Javaheri A, et al. *H. pylori* virulence factors: Influence on imune system and pathology. Mediators of Inflammation. 2014:1-9

[19] Bodger K, Bromelow K, Wyatt JJ, et al. Interleukin 10 in *Helicobacter pylori* associated gastritis: Immunohistochemical localization and in vitro effects on cytokine secretion. Journal of Clinical Pathology. 2001;**54**:285-292

[20] Lee KE, Khoi PN, Xia Y, et al. *Helicobacter pylori* and interleukin-8 in gastric cancer. World Journal of Gastroenterology. 2013;**19**:8192-8202

[21] Percival SI, Suleman L. Biofilms and *Helicobacter pylori*: Dissemination and persistence within the environment and host. World Journal of Gastrointestinal Pathophysiology. 2014;**5**:122-132

[22] Blaser MJ. Ecology of *Helicobacter pylori* in the human stomach. The Journal of Clinical Investigation. 1997;**100**:759-762

[23] Salaun L, Linz B, Suerbaum S, et al. The diversity within an expanded and redefined repertoire of phase-variable genes in *Helicobacter pylori*. Microbiology. 2004;**150**:817-830

[24] Achtman M, Azuma T, Berg DE, et al. Recombination and clonal groupings within *Helicobacter pylori* from different geographical regions. Molecular Microbiology. 1999;**32**:459-470

[25] Blaser MJ. An endangered species in the stomach. Scientific American. 2005;**292**:38-45

[26] Ahmed KS, Khan AA, Ahmed I, et al. Impact of household hygiene and water source on the prevalence and transmission of *Helicobacter pylori*: A South Indian perspective. Singapore Medical Journal. 2007;**48**:543-549

[27] Colding H, Hartzen SH, Roshanisefat H, et al. Molecular methods for typing of *Helicobacter pylori* and their applications. FEMS Immunology and Medical Microbiology. 1999;**24**:193-199

[28] van Amsterdam K, van Vliet AH, Kusters JG, et al. Of microbe and marc determinants of *Helicobacter pylori*-related diseases. FEMS Microbiology Reviews. 2006;**30**:131-156

[29] Khalifa MM, Sharaf RR, Aziz RK. *Helicobacter pylori*: A poor man's gut pathogen? Gut Pathogens. 2010;**2**:2-12

[30] Ahmed N. 23 years of the discovery of *Helicobacter pylori*: Is the debate over? Annals of Clinical Microbiology and Antimicrobials. 2005;**4**:17-19

[31] Goh KL. Prevalence of and risk factors for *Helicobacter pylori* infection in a multi-racial dyspeptic Malaysian population undergoing endoscopy. Journal of Gastroenterology and Hepatology. 1997;**12**:S29-S53

[32] Fraser AG, Scragg R, Metcalf P, et al. Prevalence of *Helicobacter pylori* infection in different ethnic groups in New Zealand children and adults. Australian and New Zealand Journal of Medicine. 1996;**26**:646-651

[33] Kawasaki M, Kawasaki T, Ogaki T, et al. Seroprevalence of *Helicobacter pylori* infection in Nepal: Low prevalence in an isolated rural village. European Journal of Gastroenterology & Hepatology. 1998;**10**:47-50

[34] de Martel C, Parsonnet J. *Helicobacter pylori* infection and gender: A meta-analysis of population-based prevalence surveys. Digestive Diseases and Sciences. 2006;**51**:2292-2301

[35] Covacci A, Telford JL, Del Giudice G, et al. *Helicobacter pylori*: Virulence and genetic geography. Science. 1998;**284**:1328-1333

[36] Vale FF, Vitor JM. Transmission pathway of *Helicobacter pylori*: Does food play a role in rural and urban areas? International Journal of Food Microbiology. 2010;**138**:1-22

[37] Bardhan PK. Epidemiological features of *Helicobacter pylori* infection in developing countries. Clinical Infectious Diseases. 1997;**25**:973-978

[38] Malaty HM, Graham DY. Importance of childhood socioeconomic status on the current prevalence of *Helicobacter pylori* infection. Gut. 1994;**35**:742-745

[39] Mendall MA, Goggin PM, Molineaux N, et al. Childhood living conditions and *Helicobacter pylori* seropositivity in adult life. Lancet. 1992;**339**:896-897

[40] Goodman KJ, Correa P, Tenganá Aux HJ, et al. *Helicobacter pylori* infection in the Colombian Andes: A population-based study of transmission pathways. American Journal of Epidemiology. 1996;**144**:290-299

[41] Peach HG, Pearce DC, Farish SJ. *Helicobacter pylori* infection in an Australian regional city: Prevalence and risk factors. The Medical Journal of Australia. 1997;**167**:310-313

[42] Kikuchi S, Kurosawa M, Sakiyama T. *Helicobacter pylori* risk associated with sibship size and family history of gastric diseases in Japanese adults. Japanese Journal of Cancer Research. 1999;**89**:1109-1112

[43] Malfertheiner P, Link A, Selgrad M. *Helicobacter pylori*: Perspectives and time trends. Nature Reviews. Gastroenterology & Hepatology. 2014;**11**:628-638

[44] Roesler BM, Zeitune JMR. Molecular epidemiology of *Helicobacter pylori* in Brazilian patients with early gastric cancer and a review to understand the prognosis of the disease. In: Roesler BM, editor. Trends in *Helicobacter pylori* Infection. Rijeka: IntechOpen; 2014

[45] Cellini L, Grande R, Di Campli E, et al. Dynamic colonization of *Helicobacter pylori* in human gastric mucosa. Scandinavian Journal of Gastroenterology. 2008:178-185

[46] Hu FZ, Ehrlich GD. Population-level virulence factors amongst pathogenic bacteria: Relation to infection outcome. Future Microbiology. 2008;**3**:31-42

[47] Cellini L. *Helicobacter pylori*: A chameleon-like approach to life. World Journal of Gastroenterology. 2014;**20**:5575-5582

[48] Goodman KJ, Correa P. The transmission of *Helicobacter pylori*. A critical review of the evidence. International Journal of Epidemiology. 1995;**24**:875-887

[49] Koffi KS, Attia KA, Adonis-Koffy LY, et al. Is the mother a risk factor for transmission of *Helicobacter pylori* infection in children between the ages of 6 months and 5 years in Còte d'Ivoire? La Medicina Tropical. 2010;**70**:359-363

[50] Weyermann M, Rothenbacher D, Brenner H. Acquisition of *Helicobacter pylori* infection in early childhood: Independent contributions of infected mother, father and siblings. The American Journal of Gastroenterology. 2009;**104**:182-189

[51] Vaira D, Holton J. Vector potential of houseflies (*Musca domestica*) for *Helicobacter pylori*. Helicobacter. 1998;**3**:65-66

[52] Grubel P, Huang L, Masubuchi N, et al. Detection of *Helicobacter pylori* DNA in houseflies (*Musca domestica*) on three continents. Lancet. 1998;**352**:788-789

[53] Junqueira ACM, Ratan A, Acerbi E, Drautz-Moses DI, Premkrishnan BNV, Costea PI, et al. The microbes of blowflies and houseflies as bacterial transmission reservoirs. Scientific Reports. 2017;**7**:16324

[54] Ho SA, Hoyle JA, Lewis FA, et al. Direct polymerase chain reaction test for detection of *Helicobacter pylori* in humans and animals. Journal of Clinical Microbiology. 1991;**29**:2543-2549

[55] Neiger R, Simpson KW. *Helicobacter* infection in dogs and cats: Facts and fiction. Journal of Veterinary Internal Medicine. 2000;**14**:125-133

[56] Momtaz H, Dabiri H, Souod N, et al. Study of *Helicobacter pylori* genotype status in cows, sheep, goats and human beings. BMC Gastroenterology. 2014;**14**:61-68

[57] Tytgat GN. Endoscopic transmission of *Helicobacter pylori*. Alimentary Pharmacology & Therapeutics. 1995;**9**(suppl. 2):105-110

[58] Peters C, Schablon A, Harling M, et al. The occupational risk of *Helicobacter pylori* infection among gastroenterologists and their assistants. BMC Infectious Diseases. 2011;**11**:154-164

[59] Mishra S, Singh V, Rao GR, et al. Detection of *Helicobacter pylori* in stool specimens: Comparative evaluation of nested PCR and antigen detection. Journal of Infection in Developing Countries. 2008;**2**:206-210

[60] Chey WD, Leontiadis G, Howden CW, Moss SF. ACG Clinical Guideline: Treatment of *Helicobacter pylori* infection. The American Journal of Gastroenterology. 2017;**112**(2): 212-239

[61] Roesler BM, Costa SCB, Zeitune JMR. Eradication treatment of *Helicobacter pylori* infection: Importance and possible relationship in preventing the development of gastric cancer. ISRN Gastroenterology. 2012:935410

[62] Manfredi M, de'Angelis GL. Eradication of *Helicobacter pylori*: In search of a better therapy. Clinical Microbiology. 2013:1-4

Chapter 2

Clinical Manifestations of the *Epsilonproteobacteria* (*Helicobacter pylori*)

Rie Louise Møller Nordestgaard,
Malene Roed Spiegelhauer, Tove Havnhøj Frandsen,
Caroline Gren, Agnes Tving Stauning and
Leif Percival Andersen

Additional information is available at the end of the chapter

http://dx.doi.org/10.5772/intechopen.80331

Abstract

Epsilonproteobacteria is a large group of Gram-negative curved or spiral microaerophilic rods, of which many are difficult to culture. Because this group of bacteria is not very well investigated, our knowledge about them is limited, and a great amount of research is still needed. At least two species are well-established human pathogens: *Campylobacter jejuni/coli* causing gastroenteritis and *Helicobacter pylori* causing gastric and extra-gastric manifestations. It is well accepted that *H. pylori* causes a chronic inflammation in the stomach and thereby causes *H. pylori*-associated gastritis, which may or may not be symptomatic. The association between *H. pylori* and peptic ulcers, MALT lymphomas, gastric cancer, idiopathic thrombocytopenic purpura, and unexplained iron-deficiency anemia (IDA) is strongly evidence based. On the other hand, pernicious (vitamin B12 deficiency) anemia, neuromyelitis optica, asthma, and Graves' disease are less evidence based. *H. pylori* may also be associated with cardiovascular disease, pancreatitis, pancreatic cancer, obesity, diabetes mellitus type 2, Parkinson's disease, liver diseases, and preeclampsia. *H. pylori* is thus involved in many gastric and extra-gastric manifestations either directly or indirectly by several proposed mechanisms including antigenic mimicry.

Keywords: *Helicobacter pylori*, infection, mimicry, gastritis, anemia, thrombocytopenic purpura, gastric cancer

1. Introduction

Epsilonproteobacteria is a large group of Gram-negative curved or spiral rods which include the genera *Campylobacter* spp., *Helicobacter* spp., *Arcobacter* spp., and *Wolinella* spp. (**Table 1**) [1]. The bacteria have microaerobic or anaerobic growth requirements, and many of these are difficult to culture from clinical samples [2]. Recent studies with identification of *Epsilonproteobacteria* by PCR have shown that these bacteria cause infections in humans more commonly than previously thought [3, 4]. The most well-known species are *Campylobacter jejuni/coli* causing gastroenteritis [2] and *Helicobacter pylori* causing gastric and extra-gastric manifestations [5].

This chapter will focus on *Helicobacter* spp. and mainly on *H. pylori*. *Helicobacter* spp. can be divided into three groups: (1) gastric *Helicobacter* spp., (2) intestinal *Helicobacter* spp., and (3) hepatobiliary *Helicobacter* spp. [6]. The knowledge about intestinal *Helicobacter* spp. in human diseases is very limited mainly because they are very difficult to culture. In contrast to the intestinal and hepatobiliary *Helicobacter* spp., the gastric *Helicobacter* spp. produce a great amount of urease, which is important for its survival in the stomach by neutralizing acid, thereby creating a neutral microenvironment [7]. Urease is also crucial for the bacteria's survival through antigenic shedding where urease captures human antibodies [8]. The human gastric *Helicobacter* sp., *H. pylori*, is the most intensively investigated *Helicobacter* sp., but gastric *Helicobacter* spp. from animals (*Helicobacter heilmannii*, *Helicobacter bizzozeronii*, *Helicobacter suis*, etc.) have also been found in the human stomach [9]. These bacteria colonize the stomach in very different ways. *H. pylori* colonizes the antrum part of the stomach on the surface between epithelial cells and can actively move down between the epithelial cells [10]. On the other hand, *Helicobacter* sp. from animals colonizes the parietal cell glands in the corpus/fundus part of the stomach which may contribute to other manifestations than those caused by *H. pylori* [11]. Usually, a stronger cellular immune response is seen in *H. pylori* in comparison to the animal-associated *Helicobacter* spp. [11].

Genus	Species
Arcobacter	*anaerophilus, aquimarinus, bivalviorum, butzleri, cibarius, cloacae, cryaerophilus, defluvii, ebronensis, ellisii, haliotis, halophilus, lanthieri, lekithochrous, marinus, molluscorum, mythili, nitrofigilis, pacificus, suis, thereius, trophiarum, venerupis*
Campylobacter	*avium, canadensis, coli, concisus, corcagiensis, cuniculorum, curvus, fetus* subsp. *fetus, fetus* subsp. *testudinum, fetus* subsp. *venerealis, geochelonis, helveticus, hepaticus, hominis, hyoilei, hyointestinalis* subsp. *hyointestinalis, hyointestinalis* subsp. *lawsonii, iguanorium, insulaenigrae, jejuni* subsp. *doylei, jejuni* subsp. *jejuni, lanienae, lari* subsp. *concheus, lari* subsp. *lari, mucosalis, ornithocola, pyloridis, pinnipediorum, pinnipediorum* subsp. *caledonicus, pinnipediorum* subsp. *pinnipediorum, rectus, showae, sputorum, subantarcticus, upsaliensis, ureolyticus, volucris*
Helicobacter	*acinonychis, ailurogastricus, anseris, apri, aurati, baculiformis, bilis, bizzozeronii, brantae, canadensis, canicola, canis, cetorum, cholecystus, cinaedi, cynogastricus, equorum, felis, fennelliae, ganmani, heilmannii, hepaticus, himalayensis, jaachi, japonicus, macacae, marmotae, mastomyrinus, mesocricetorum, muridarum, mustelae, pamatensis, pullorum, pylori, rodentium, salomonis, saguini, suis, trogontum, typhlonius, valdiviensis*
Wolinella	*succinogenes*

Table 1. The species belonging to the four largest groups of ***Epsilonproteobacteria*** [102].

H. pylori may either cause direct or indirect damage to the stomach: direct damage where *H. pylori* infections disintegrate gastric mucosa and cause apoptosis through cytotoxin-associated gene A (CagA) and vacuolating toxin (VacA) or indirect damage where *H. pylori* induces a strong and chronic immune response by activating B and T lymphocytes, macrophages, neutrophilic lymphocytes, and probably also eosinophil leukocytes. T cell-activated B lymphocytes, regulatory T cells (Treg), and T helper 17 cells (Th17) are some of the B and T lymphocytes that are important in *H. pylori* infections. T cell-activated B lymphocytes are responsible for a strong humoral immune response primarily toward *H. pylori* urease, flagella, CagA, and VacA. These activated B and T lymphocytes release a large range of cytokines of which IL1-β, TNF-α, INF-γ, IL6, IL-8, IL-10, IL-17, and cyclooxygenase-2 (COX-2) are the most important cytokines in severe *H. pylori* infections [12, 13].

Many microorganisms can cause autoimmune diseases. The mechanisms involved include molecular mimicry (when bacterial antigens cross-react with human tissue), epitope spreading, bystander effect, microbial superantigens, immune complex formation, MHC class II expression on nonimmune cells, and high levels of pro-inflammatory cytokines [14–17]. *H. pylori* has been implicated in both organ-specific and non-organ-specific autoimmune diseases and has been investigated sporadically or systematically in 95 autoimmune-related diseases [18]. Many mechanisms underlying the antigenic mimicry between *H. pylori* and the host have been proposed. Efforts have been made to identify homologous sequences between *H. pylori* and host polypeptides. H+/K + −adenosine triphosphatase, Lewis antigens, and lipopolysaccharide seem to be autoantigens in autoimmune gastritis. Glycoproteins and Lewis antigens may be autoantigens directed against platelets in idiopathic thrombocytopenic purpura (ITP). Lewis antigens, heat shock protein 60 (HSP60), and 160/180 kDa antigens appear to be autoantigens to the endothelium, while alpha-carbon anhydrase and plasminogen-binding proteins could to be autoantigens in the pancreas [13].

All in all, *H. pylori* can cause both gastric and extra-gastric diseases through a complex mechanism involving both host and bacterial factors.

2. Gastritis and peptic ulcer

Whenever *H. pylori* is found in the human stomach, there is never just a simple colonization. Instead, there is always a cellular and humoral immune response confirming that *H. pylori* causes infection [10, 19, 20]. Thus, patients with gastritis and *H. pylori* have *H. pylori*-related gastritis. However, if there is no *H. pylori* infection, patients may have functional gastritis but no inflammation. *H. pylori*-related gastritis may benefit from antibiotic treatment, whereas there is no indication for antibiotic treatment for functional gastritis [21].

Peptic ulcers occur in about 10% of patients infected with *H. pylori* where most (80%) are duodenal ulcers [19]. More than 90% of duodenal ulcers are caused by *H. pylori* [19]. The pathogenesis of these ulcers is not clear, but they often occur in the part of the duodenum where the flow from the stomach content is the highest. Duodenal ulcers may be caused by a combination of physical, physiological, and immunologic effects as well as *H. pylori*. Patients with duodenal ulcers almost always benefit from antibiotic treatment. More than 60% of gastric ulcers are caused by *H. pylori,* while the remaining 40% may be caused by different sources such as medication (NSAID, etc.) [21, 22]. Gastric ulcers are often found in the isthmus area of the stomach where the amount of blood flow of the stomach is the lowest. *H. pylori* stimulates the production of platelet-activating factor (PAF) which acts on angiogenesis by contracting blood vessels [23]. *H. pylori* has a direct damaging effect on the epithelium and interferes with the immune system in many ways [24]. However, the mechanisms are very complex, and the pathogenesis is still not completely understood.

3. Mucosa-associated lymphoid tissue (MALT) lymphomas

MALT lymphomas are a group of lymphomas which arise in the tissue normally devoid of lymphoid tissue, such as the stomach. These tissues accumulate lymphoid tissue during chronic antigenic stimulation such as chronic infections and autoimmune diseases. *H. pylori* causes about 80% of low-grade MALT lymphomas and 60% of high-grade MALT lymphomas [19]. Eradication of *H. pylori* stops the progression in most cases, and 60–80% of early-state low-grade MALT lymphomas will regress [25]. The mechanism by which *H. pylori* induces MALT lymphomas is unclear, and there is no evident correlation between MALT lymphomas and *H. pylori* virulence factors [26]. One theory is that the development of gastric MALT lymphomas in patients with *H. pylori* could be secondary to chronic antigenic stimulation of the immune system by the pathogen [27]. However, as in many other diseases, antigenic mimicry may also play a role [27]. Finally, it is possible that MALT lymphomas are correlated to non-pylori *Helicobacter* spp. instead of *H. pylori* [28, 29].

4. Gastric cancer

H. pylori causes approximately 80% of all gastric cancer cases, and in 1994 *H. pylori* became categorized as a Group 1 carcinogen meaning that *H. pylori* is a definite carcinogen to humans [30].

The development of gastric cancer is a complex process that depends on *H. pylori* virulence factors, host mucosa properties, immunological reactions to infections, as well as environmental factors in the stomach. In *H. pylori,* virulence factors like CagA and VacA have been suggested to influence cancer development. *CagA* gene and the type IV secretion system (T4SS) are encoded by a 40-kb DNA fragment called *cag* pathogenicity island (*cag*PAI) [19, 31]. CagA protein infects host gastric epithelial cells via the T4SS, where it is tyrosine-phosphorylated by host kinases at specific glutamate-proline-isoleucine-tyrosine-alanine (EPIYA) motifs [31, 32]. CagA thereafter interferes with different host cell-signaling pathways causing changes in cell growth, polarity, and motility, thereby increasing the risk for gastric cancer [19, 32]. VacA toxin affects gastric epithelial cells in a similar manner by affecting the host's inflammatory response as well as cellular apoptosis among other ways [19]. Other host factors could be high-salt diets and iron deficiency, which have been proven to increase the risk for gastric cancer [33, 34].

If *H. pylori* is treated in the early premalignant stages (atrophic gastritis), further cancer development can be prevented [35]. If intestinal metaplasia has developed, it is believed that antibiotic treatment has no effect [21]. As with gastritis and peptic ulcers, the relationship between *H. pylori* and gastric cancer has many loose ends that need to be explained before we can completely understand the process.

5. Idiopathic thrombocytopenic purpura (ITP)

Idiopathic thrombocytopenic purpura or immune thrombocytopenic purpura (ITP) is an acquired autoimmune disease resulting in the destruction of antibody-covered platelets and decreased platelet production. This results in an increased risk for bruising and bleeding. ITP is defined as a platelet count $<100 \times 10^9$ /L, may be either primary or secondary, and is classified as acute, persistent, or chronic [36].

The mechanism that leads to ITP in *H. pylori*-infected patients is not entirely established. It is proposed that molecular mimicry may be involved [13]. Cross-reactivity between platelet-associated immunoglobulin G and CagA has been found, which suggests that mimicry through CagA may play a role in the development of ITP [37].

It is well established that *H. pylori* screening may be warranted in patients with ITP. A systematic review from 2009 with 696 evaluable patients found that in patients with *H. pylori* infection, eradication of the bacteria led to a complete treatment response in 43% of the patients and an overall response (platelet count $\geq 30 \times 10^9$/L and at least a doubling of initial platelet count) of 50%. The treatment tended to be more effective in milder forms of thrombocytopenia. The authors found that the predictors of treatment response were quite heterogeneous from study to study. Shorter duration of ITP was consistently found, and response rates tended to be higher in countries with a higher prevalence of *H. pylori* [38]. In the highly *H. pylori* prevalent country of South Korea, a more recent prospective study with 26 patients with persistent or chronic ITP investigated the efficacy of *H. pylori* eradication as a first-line treatment in patients with moderate thrombocytopenia [39]. The study found an eradication rate of 80% and a maximal complete response rate of 65% [39].

The most recent ITP guidelines from the American Society of Hematology (ASH) recommend eradication therapy in adult ITP patients with *H. pylori* infection. They do not define which patients should be screened or at what point in the course of the illness patients should receive treatment [36]. ASH recommends against routine testing in children because of diverging results but rather argues for the consultation with a pediatric gastroenterologist beforehand. Since the publication of the ASH guidelines, a randomized-controlled trial (RCT) with 85 ITP-affected children has been published. Twenty-two children were *H. pylori* infected, and they were randomized to receive either eradication therapy or no therapy. Complete response was achieved in 60% of the treated children compared to 18% of the children who were not treated. The authors suggested that *H. pylori* infection may play a bigger role in the pediatric ITP population than the earlier notions. It is also noted that 86% of the patients had CagA antibodies and 82% harbored VacA antibodies [40]. The recently updated joint European Society for Pediatric Gastroenterology, Hepatology, and Nutrition/North American Society for Pediatric Gastroenterology, Hepatology, and Nutrition (ESPGHAN/NASPGHAN) guidelines recommend testing for *H. pylori* in children with chronic ITP [41].

6. Iron-deficiency anemia

H. pylori infection has also been linked to iron-deficiency anemia (IDA) [42–44]. Mechanisms that cause IDA may increase iron loss due to hemorrhagic gastritis, gastric cancer, peptic ulcers, iron utilization for bacterial growth, achlorhydria resulting in reduced iron uptake, and reduced secretion of ascorbic acid [45].

A meta-analysis comprising 15,183 patients from 20 studies found an association between *H. pylori* infection and IDA (odds ratio (OR) 2.22) [46]. They also found a greater effect of eradication therapy plus iron than iron supplements alone but with heterogeneous results. Adult IDA patients reacted more strongly to eradication than children and adolescents, and bismuth triple therapy seemed to be more effective than proton pump inhibitor (PPI) triple therapy. The authors do not recommend a population-based screening for *H. pylori* to prevent IDA [46].

On the other hand, Herschko et al. studied 160 patients with autoimmune gastritis, of whom 83 presented with IDA [47]. When stratifying by age, they found a decreasing prevalence of coexistent *H. pylori* infection with increasing age: 88% at age <20 years, 47% at 20–40 years, 38% at 41–60 years, and 13% at age >60 years. A possible explanation, which other authors also have mentioned, is that *H. pylori* demands an acidic environment to survive, which no longer exists in advanced atrophic anemia. This might suggest that *H. pylori* infection in autoimmune gastritis may represent an early phase of the disease in which an infectious process is gradually replaced by an autoimmune disease terminating in a burned-out infection and the irreversible destruction of gastric mucosa. This might explain why younger patients with IDA have a high prevalence of *H. pylori* infection [47].

The British Society of Gastroenterology recommends noninvasive testing and antibiotic treatment for *H. pylori* in patients with IDA and normal esophagogastroduodenoscopy and colonoscopy [48]. The American College of Gastroenterology also recommends testing for *H. pylori* in patients with unexplained IDA [49]. The association between IDA and *H. pylori* infection in the pediatric population is less studied and with heterogeneous results. ESPGHAN/NASPGHAN guidelines propose that in children with refractory IDA where there is an indication for upper endoscopy, it might be considered taking biopsies to test for *H. pylori* [41].

7. Vitamin B_{12} deficiency anemia

Vitamin B_{12} (cobalamin) deficiency is estimated to affect approximately 10–15% of the population older than 60 years. There are several causes where pernicious anemia and food-cobalamin malabsorption are the most common reasons. Cobalamin is obtained primarily from food through a complicated process where an acidic environment releases cobalamin from food and thereafter binds to intrinsic factors secreted from parietal cells and finally is absorbed by specific receptors in the terminal ileum. Pernicious anemia is an autoimmune disorder consisting of chronic atrophic gastritis, decreased acid secretion, and antibodies directed against parietal cells and/or intrinsic factors, thereby leading to decreased cobalamin absorption. *H. pylori* possibly stimulates these antibodies directed against parietal cells/intrinsic factors, thereby inducing pernicious anemia. In food-cobalamin malabsorption, there is an inability to absorb food-bound or protein-bound cobalamin in a person that normally can absorb free cobalamin. *H. pylori* infection predisposes to a more severe form of food-cobalamin malabsorption [50].

As mentioned above, it has been proposed that B_{12} deficiency can arise as the result of a late phase of *H. pylori*-induced atrophic gastritis [47]. This theory has been mentioned already in the early 1990s [51]. In a prospective case series with 138 patients with megaloblastic anemia and low cobalamin, it was found that 56% had *H. pylori* infection. Eradication therapy was successful in 40% of the infected patients, and the hematological parameters and B_{12} levels improved in all these patients without complementary cobalamin therapy [52].

The literature regarding the association between *H. pylori* and pernicious anemia shows more heterogeneous results than for ITP and IDA [52]. Therefore, treatment guidelines do not yet recommend screening for *H. pylori* in pernicious anemia. However, the Maastricht V/Florence Consensus Report does recommend that in all three of the abovementioned disorders *H. pylori* should be screened for and eradicated [21].

8. Cardiovascular disease

Studies indicate an association between *H. pylori* and cardiovascular disease (CVD) [53, 54]. However, the stratification of patient groups and methods are very heterogeneous which may be the reason for the very diverging results in the studies [53]. *H. pylori* seems to mostly be associated with coronary atherosclerosis [55, 56]. This is in accordance with an unpublished study where we found increased antibodies to *H. pylori*, but not to *Chlamydophila pneumoniae* and *Cytomegalovirus* in patients undergoing surgery for coronary atherosclerosis. *H. pylori* can survive in monocytes, and it might be speculated whether the bacteria could be transferred from the stomach to the coronary vessels. Here, *H. pylori* may stimulate PAF and other factors that may act on angiogenesis [23, 56]. *H. pylori* may also stimulate the atherogenesis through molecular mimicry or vitamin B12 and folate malabsorption [13, 53, 54]. In addition, *H. pylori* may change the lipid profile by increasing LDL levels and decreasing HDL levels as seen in many other infections, which leads to atherogenesis [53, 54, 57–59].

9. Pancreatitis and pancreatic cancer

Studies have shown a correlation between increased antibody levels to *H. pylori* in patients with pancreatitis and pancreatic cancer [60–63]. In an unpublished study, we showed that in more than 50% of patients with pancreatitis *H. pylori* was cultured from the antral part of the stomach. The interaction leading to pancreatic cancer is unknown, but *H. pylori* infection in the antral part of the stomach decreases the production of somatostatin. This increases pancreatic bicarbonate and secretin which stimulates ductal epithelial cell proliferation [64]. In addition, studies indicate that *H. pylori* increases the risk of autoimmune pancreatitis through molecular mimicry and thereby increases the risk for pancreatic cancer [13, 60, 63–65]. These findings are of great interest and need further intensive research.

10. Obesity and diabetes mellitus type 2

Obesity is becoming a worldwide problem, and population studies have shown that in the same areas where the prevalence of *H. pylori* is decreasing, the prevalence of obesity is increasing [21, 66]. An implication of obesity could be diabetes mellitus type 2. A possible mechanism in which *H. pylori* affects obesity and thereby also affects type 2 diabetes is persistent damage of gastric mucosa, e.g., chronic gastritis. This might affect ghrelin production, thereby changing food intake and increasing body weight [67, 68].

Ghrelin is a hormone mainly produced by endocrine cells in the gastrointestinal mucosa and is released to the surroundings. This molecule is important for stimulating food intake and weight gain [69]. The damages that *H. pylori* introduce on gastric mucosa reduce the number of ghrelin-producing cells and decrease plasma ghrelin concentrations significantly, thereby reducing the feeling of satiety which can lead to obesity [67, 68, 70].

Ghrelin also seems to play a role in fat metabolism and glucose homeostasis, which can lead to a cross-reaction between lipid and glucose metabolisms that may result in insulin resistance [71]. However, one thing is clear, diabetes mellitus type 2 is a multifactorial disease, and *H. pylori* is only one of the many risk factors. *H. pylori* may also act on leptin or by activating cytokines that together can have an effect on insulin secretion [72, 73].

Although many studies have shown that there could be a correlation between *H. pylori* and obesity and diabetes mellitus type 2, other studies have shown that there are none and the correlation is still uncertain [66, 74].

11. Parkinson's and Alzheimer's diseases

Numerous studies indicate that *H. pylori* infection is associated with a more rapid development of cognitive and functional deterioration. Furthermore, eradication of *H. pylori* could give an improved disease severity [75–78]. Also, a study by Weller et al. showed that the presence of CagA antibodies is associated with a poorer Parkinson's prognosis [79]. It is proposed that *H. pylori* initiates the destruction of mitochondria and together with antigenic mimicry stimulates Parkinson's disease [72]. Only few studies focus on *H. pylori* and Alzheimer's disease, and they are too preliminary to show a causal or therapeutic association [72, 75].

12. Neuromyelitis optica

Several studies have shown a correlation between *H. pylori* and neuromyelitis optica (NMO) [18]. NMO is a disease where antibodies attack aquaporin-4 on astrocytes in the central nervous system [80]. There is a close relationship between *H. pylori* and antibodies to aquaporin-4, and thus molecular mimicry could play a role [18].

13. Asthma

The prevalence of asthma is increasing in areas where the prevalence of *H. pylori* is decreasing [81]. Meta-analyses have found an inverse correlation between *H. pylori* and asthma, but the mechanism is unclear [72, 82, 83]. CagA-positive *H. pylori* strains especially have been found to have a greater inverse relationship with asthma than those without *H. pylori* [81]. The long-established hygiene hypothesis, where a lack of exposure to infectious agents leads to an increased risk for allergens, has been proposed as one way in which an absence of *H. pylori* causes asthma [82]. Th2-mediated immune responses drive allergies, while Th1-mediated immune responses inhibit these reactions. *H. pylori* appears to stimulate Th1-mediated immune responses but inhibit Th2-mediated immune responses through neutrophil-activating protein (HP-NAP), thereby inhibiting asthma development [84]. Another possible mechanism of *H. pylori* is upregulation of Treg cells which can control Th2-mediated immune responses [82]. A mouse study by Arnold et al. proved that *H. pylori* infection protected mice against asthma and an upregulation of Treg cells was found in mice infected with *H. pylori* [85]. Thus, *H. pylori* could inhibit asthma in a multitude of ways.

14. Hepatobiliary diseases

Non-pylori *Helicobacter* species have been isolated from the liver of a variety of animals. *H. hepaticus*, *H. bilis*, and *H. cholecystus* are involved in the pathogenesis of chronic liver diseases and liver carcinomas [86–88]. *H. pylori*, *H. hepaticus*, *H. bilis*, and *H. cholecystus* have been detected in the human hepatobiliary tissue mainly by PCR [89–91]. Several studies have shown an increased prevalence of *H. pylori* in patients with hepatocellular carcinomas (HCC), liver encephalopathy (HE), liver fibrosis, cholangiocarcinoma (CCA), primary biliary cirrhosis (PBC), and primary sclerosing cholangitis [92]. Much interest has been linked to HCC and CCA which histologically is characterized as adenocarcinomas. The pathogenesis has been proposed to follow the same pattern as in stomach cancer: hyperplasia, metaplasia, dysplasia, and lastly cancer [92]. Inflammatory cytokines and chemokines may play an important role in the pathogenesis. HE is a frequent complication to liver cirrhosis with a wide variety of neuropsychiatric symptoms, and high levels of ammonia play an important role in the pathogenesis [93]. *H. pylori* produces urease which reacts to ammonium, which might explain a possible mechanism in HE development. Liver fibrosis, among other ways, may be caused by *H. pylori* stimulating hepatocytes and results in accumulation of collagen, thereby causing fibrosis [63]. Some of the risk factors for these cancers are population genetics, geographical and environmental factors, cholelithiasis, obesity, chronic inflammation, and obstruction of the bile duct [92, 94].

15. Autoimmune thyroid diseases

Both Graves' disease and Hashimoto's thyroiditis are autoimmune diseases in the thyroid. Graves' disease is characterized by hyperthyroidism and an enlarged gland, while Hashimoto's thyroiditis is characterized by hypothyroidism and the destruction of thyroid tissue. There is an association between Graves' disease and *H. pylori,* where CagA is most likely an important virulence factor [95]. A study by Bassi et al. showed that 82% (43/52) of patients with Graves' disease were positive for *H. pylori,* where 84% (36/43) of *H. pylori*-positive Graves' disease patients were positive for CagA antigens. Also, a different study by Bertalot et al. showed a reduction in thyroid autoantibodies following *H. pylori* eradication [96]. Amino acid sequences of thyroid peroxidase and CagA are very similar, and cross-reactivity is a possible mechanism by which *H. pylori* increases the risk of developing Graves' disease [18, 95]. In addition, Graves' disease is often found with other autoimmune diseases which may reflect the ability of *H. pylori* to induce multiple autoimmune diseases simultaneously [97]. However, the same cannot be said about Hashimoto's thyroiditis where a significant association between Hashimoto's thyroiditis and *H. pylori* was not found by Bassi et al. [95].

16. Preeclampsia

The first study investigating the association between *H. pylori* infection and preeclampsia (PE) was conducted in Italy and published in 2006 [98]. It was found that 32% of women with a normal pregnancy harbored anti-*H. pylori* antibodies compared to 51% of preeclamptic women. The difference was even bigger when looking at the presence of anti-CagA antibodies: 15 vs. 81% in women with a normal pregnancy vs. preeclamptic women. The authors concluded that the increased inflammatory activity in *H. pylori*-infected patients may contribute to the development of PE, especially in CagA strains. Interestingly, no *H. pylori* DNA was present in the placentas that were studied, and therefore the inflammation is probably not locally induced.

A review from 2014 concluded that there is evidence indicating that *H. pylori* negatively influences human reproductivity, including PE [99]. This is probably due to both increased inflammatory activity and antigenic mimicry with CagA-positive strains appearing to be the most important culprits [99]. A recent meta-analysis of observational studies with 9787 women (879 preeclamptic) confirmed these theories, with an OR of 2.32 for anti-*H. pylori* antibodies in cases compared to controls and an OR of 3.97 for having anti-CagA antibodies in preeclamptic patients [100]. A review on the topic of infections and the risk of PE mentions *H. pylori* as a possible cause of PE and recommends that screening (and treatment) of known infectious organisms causing PE should be included in antenatal programs [101]. However, as mentioned by Bellos et al., it is yet unknown if *H. pylori* predisposes to mild or severe PE, at which gestational age optimal screening should be conducted, and most importantly how effective eradiation is in terms of reducing the incidence and severity of PE [100].

17. Discussion

H. pylori can induce many pathogenic reactions in infected individuals. There are mainly three different ways *H. pylori* acts. (1) The bacteria have several virulence factors (Cag PAI, Vac A, etc.) that can cause direct damage and apoptosis of epithelial cells in the stomach and can stimulate mast cells to liberate PAF which affects the angiogenesis in the stomach. This may be some of the main actions on gastric diseases such as peptic ulcers and gastric cancer (**Figure 1**). (2) There is a strong cellular and humoral immune response to *H. pylori* with the release of different cytokines and chemokines. Cytokines and chemokines subsequently react both in the stomach and in extra-gastric organs (**Figure 2**). In addition, several *H. pylori* antigens are structurally like antigens of the human body and therefore may cause cross-reactions (antigenic mimicry) (**Figure 3**). All these pathogenic mechanisms of *H. pylori* may result in different diseases both in the stomach and in extra-gastric organs.

The role of *H. pylori* in relation to gastritis, peptic ulcers, MALT lymphomas, and gastric cancer is well known and established. However, there is confusion about the difference between

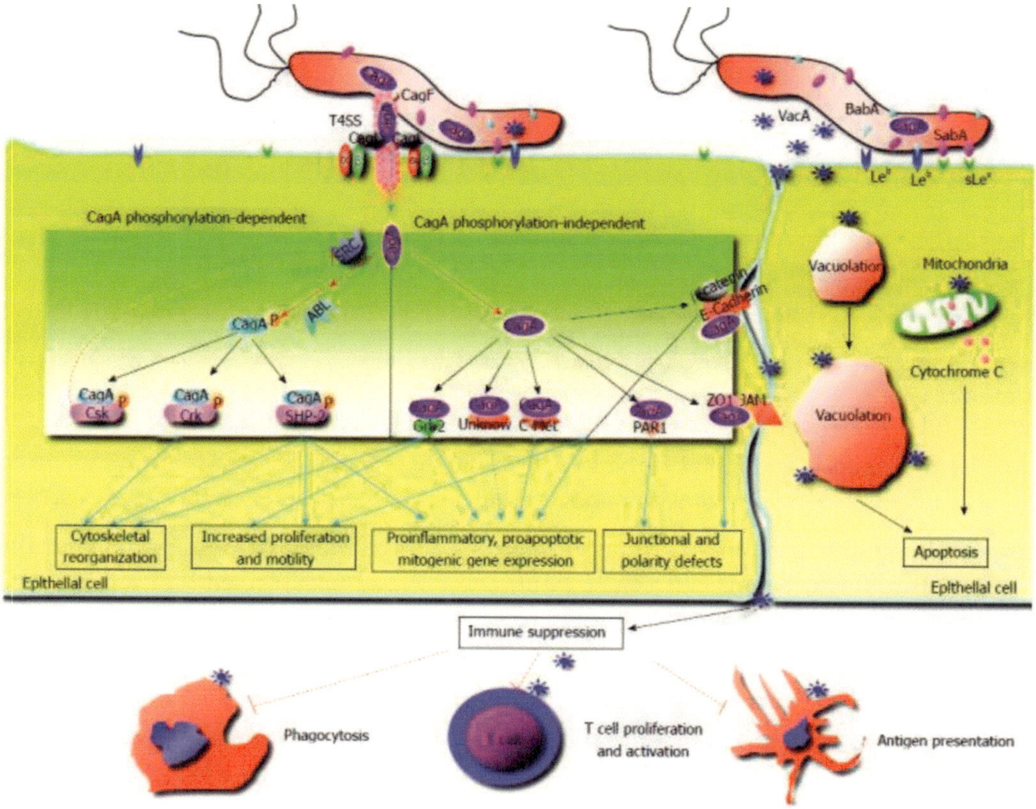

Figure 1. The roles of the main virulence factors in pathogenesis of *Helicobacter pylori* infection [6]. Adherence of *Helicobacter pylori* to gastric epithelial cells is mediated by BabA and SabA binding Leb and Lewis x/a, respectively. CagA is translocated into epithelial cells through T4SS and then tyrosine-phosphorylated at EPIYA sites by Src and Abl kinases. CagA contributes to alteration of myriad signaling transduction, which affects host cell physiology with disruption of intercellular junctions, loss of cell polarity, promotion of inflammation, dysregulation of cellular apoptosis, and proliferation. VacA inducts cytoplasmic vacuolation, apoptosis, and immune suppression [6, 103].

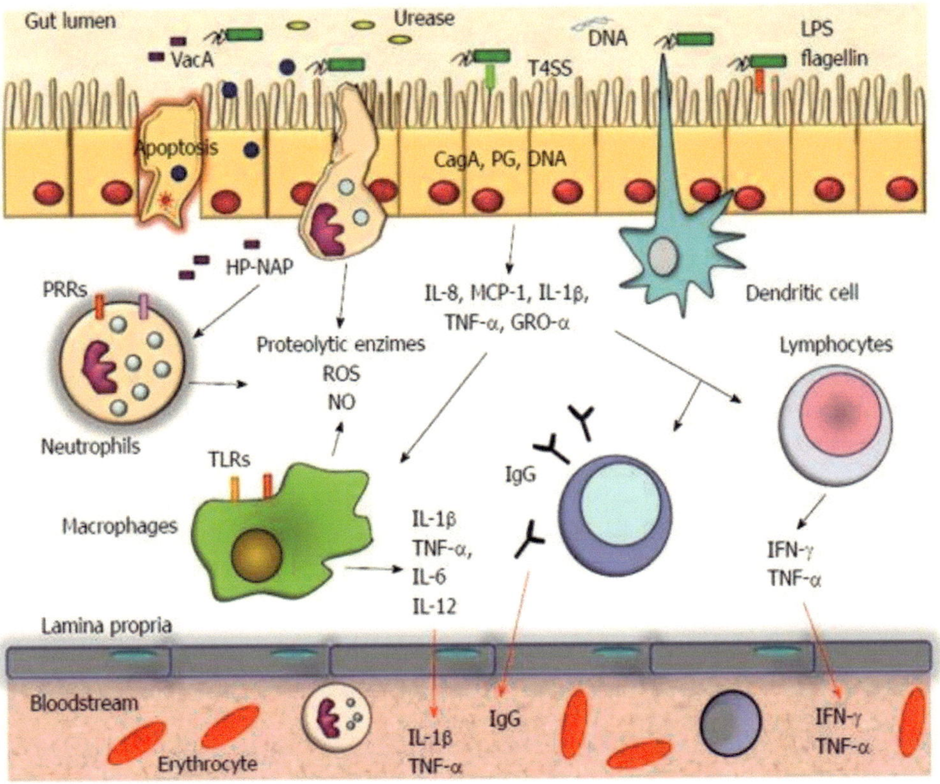

Figure 2. The inflammatory response in *Helicobacter pylori* infection. Immune cells are recruited to the lamina propria of the gastric epithelium by chemokines and cytokines (IL-8, MCP-1, GRO-α, IL-1β, TNF-α) produced by epithelial cells or directly by bacterial products including *H. pylori* neutrophil-activating protein, VacA, and urease. At the site of infection, the immune cells are activated and exert their effector functions, including the production of cytokines (IL-1β, TNF-α, IL-6, IL-12, IFN-γ), chemokines (IL-8, MCP-1), proteolytic enzymes, oxide nitric (NO), and reactive oxygen species (ROS). PG, peptidoglycan; T4SS, type IV secretion system; IL, interleukin; TNF, tumor necrosis factor; MCP, macrophage chemotactic protein; GRO, growth-regulated oncogene [104].

functional dyspepsia and *H. pylori*-induced gastritis even though *H. pylori* is always followed by a strong cellular and humoral immune response and fulfills the criteria for a true infection.

As with many other infections, *H. pylori* infection does not always cause symptoms. The evidence-based associations between *H. pylori* and ITP and unexplained IDA are less well known. Patients with these diseases should be tested for *H. pylori*. There are slightly weaker associations found between *H. pylori* and B_{12} deficiency anemia, neuromyelitis optica, and Graves' disease, and patients with these diseases should also be tested for *H. pylori* [21].

Weaker associations between *H. pylori* and cardiovascular disease, pancreatic cancer, pancreatitis, obesity and type 2 diabetes, Parkinson's disease, asthma, liver diseases, and preeclampsia have been found. *H. pylori* possibly causes these diseases through antigenic mimicry, and affected patients should be considered for *H. pylori* testing.

In conclusion, a variety of diseases may be caused by *H. pylori*, and affected patients should be tested for *H. pylori*. However, further larger and more well-designed studies with better stratification of patients and better diagnostics of *H. pylori* are needed.

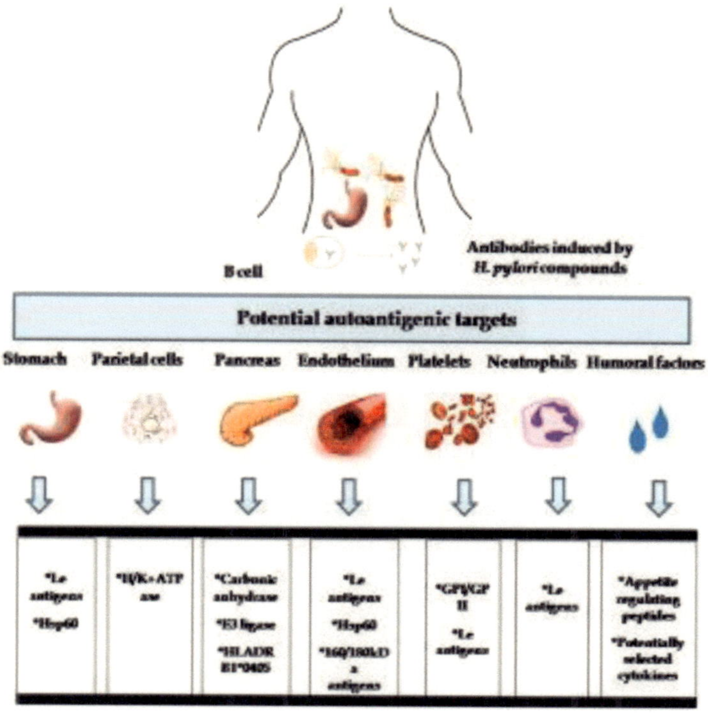

Figure 3. Hypothesis of autoimmune disorders due to molecular mimicry between *Helicobacter pylori* and the host components. Chronic exposure of the host immune system to *Helicobacter pylori* (*H. pylori*) components that have homologous sequences with the host cellular or soluble compounds may initiate the production of autoantibodies. However, how often the autoantibodies arising during *H. pylori* infection are involved in various post-infectious pathologies should be elucidated. The graph shows the examples of host targets for the antibodies induced by *H. pylori* components. GP, glycoproteins; HSP, heat shock protein; H+/K+ ATPase, H+/K + –adenosine triphosphatase; HLA, human leukocyte antigens; CCRL1, CC chemokine receptor-like 1; Le, Lewis antigens [105].

18. Conclusion

A variety of diseases are may be caused by ***H. pylori***; some such as peptic ulcer and gastric cancer by a direct effect on the gastric epithelial cells cause cell damage and apoptosis. The complex immune response to ***H. pylori*** contributes to the pathogenesis such as mast cells liberating PAF which affect the angiogenesis in the stomach. The complex immune response to ***H. pylori*** is also involved in the pathogenesis of extra-gastric manifestations of ***H. pylori*** infection. In addition to the immune response to ***H. pylori***, ***H. pylori*** also contains a lot of antigens which cross-react with human antigens (antigenic mimicry) that is responsible for many autoimmune diseases such as thrombocytopenia purpura, B12 deficiency anemia, neuromyelitis optica, Graves' disease, etc. Thus, ***H. pylori*** causes or may cause a lot of well-known and less well-investigated diseases, and these patients should be tested for ***H. pylori***. However, many of these diseases are rather rare especially in children that need larger, and more well-designed multicenter studies with better stratification of patients and better diagnostics of ***H. pylori*** for prober studies are needed. In addition, little is known about the exact virulence and pathogenic mechanisms of *H. pylori*, and basic research in these diseases is urgently needed.

Conflict of interest

The authors declare that they have no conflict of interest.

Abbreviations

CVD	cardiovascular disease
CagA	cytotoxin-associated gene A
H. pylori	*Helicobacter pylori*
IDA	iron-deficiency anemia
ITP	idiopathic thrombocytopenic purpura
MALT	mucosa-associated lymphoid tissue
NMO	neuromyelitis optica
OR	odds ratio
PAF	platelet-activating factor
PE	preeclampsia
Treg	regulatory T cells
VacA	vacuolating toxin

Author details

Rie Louise Møller Nordestgaard, Malene Roed Spiegelhauer, Tove Havnhøj Frandsen, Caroline Gren, Agnes Tving Stauning and Leif Percival Andersen*

*Address all correspondence to: leif.percival.andersen@regionh.dk

The Helicobacter Research Center, Department of Clinical Microbiology 9301, Copenhagen University Hospital (Rigshospitalet), Copenhagen, Denmark

References

[1] Campbell BJ, Engel AS, Porter ML, et al. The versatile ε-proteobacteria: Key players in sulphidic habitats. Nature Reviews. Microbiology. 2006;**4**(6):458-468

[2] Engberg J, On SL, Harrington CS, et al. Prevalence of *Campylobacter, Arcobacter, Helicobacter,* and *Sutterella spp.* in human fecal samples as estimated by a reevaluation of isolation methods for Campylobacters. Journal of Clinical Microbiology. 2000;**38**(1):286-291

[3] Cornelius AJ, Chambers S, Aitken J, et al. Epsilonproteobacteria in humans, New Zealand. Emerging Infectious Diseases. 2012;**18**(3):510-512

[4] Maher M, Finnegan C, Collins E, et al. Evaluation of culture methods and a DNA probe-based PCR assay for detection of *Campylobacter* species in clinical specimens of feces. Journal of Clinical Microbiology. 2003;**41**(7):2980-2986

[5] Cover TL, Blaser MJ. *Helicobacter pylori* in health and disease. Gastroenterology. 2009; **136**(6):1863-1873

[6] Andersen LP. New *Helicobacter* species in humans. Digestive Diseases. 2001;**19**(2):112-115

[7] Andersen LP. Colonization and infection by *Helicobacter pylori* in Humans. Helicobacter. 2007;**12**(s2):12-15

[8] Mobley HLT. *Helicobacter pylori* Factors associated with disease development. Gastroenterology. 1997;**113**(6):S21-S28

[9] Andersen LP, Boye K, Blom J, et al. Characterization of a culturable "*Gastrospirillum hominis*" (*Helicobacter heilmannii*) strain isolated from human gastric mucosa. Journal of Clinical Microbiology. 1999;**37**(4):1069-1076

[10] Kim N. *Helicobacter pylori*. Springer Singapore: Singapore; 2016

[11] Holck S, Ingeholm P, Blom J, et al. The histopathology of human gastric mucosa inhabited by *Helicobacter heilmannii* -like (*Gastrospirillum hominis*) organisms, including the first culturable case. Acta Pathologica, Microbiologica, et Immunologica Scandinavica. 1997; **105**(7-12):746-756

[12] Kobayashi M, Lee H, Nakayama J, et al. Roles of gastric mucin-type O-glycans in the pathogenesis of *Helicobacter pylori* infection. Glycobiology. 2009;**19**(5):453-461

[13] Chmiela M, Gonciarz W. Molecular mimicry in *Helicobacter pylori* infections. World Journal of Gastroenterology. 2017;**23**(22):3964-3977

[14] Vial T, Descotes J. Autoimmune diseases and vaccinations. European Journal of Dermatology. 2004;**14**(2):86-90

[15] McCoy L, Tsunoda I, Fujinami RS. Multiple sclerosis and virus induced immune responses: Autoimmunity can be primed by molecular mimicry and augmented by bystander activation. Autoimmunity. 2006;**39**(1):9-19

[16] Ram M, Shoenfeld Y. Hepatitis B: Infection, vaccination and autoimmunity. IMAJ. 2008; **10**:61-64

[17] Ravel G, Christ M, Horand F, et al. Autoimmunity, environmental exposure and vaccination: Is there a link? Toxicology. 2004;**196**(3):211-216

[18] Smyk DS, Koutsoumpas AL, Mytilinaiou MG, et al. *Helicobacter pylori* and autoimmune disease: Cause or bystander. World Journal of Gastroenterology. 2014;**20**(3):613-629

[19] Hagymási K, Tulassay Z. *Helicobacter pylori* infection: New pathogenetic and clinical aspects. World Journal of Gastroenterology. 2014;**20**(21):6386-6399

[20] Malfertheiner P, Michetti P, Price A. *Helicobacter pylori*: An Atlas. 1st ed. Science Press; 1996

[21] Malfertheiner P, Megraud F, Morain CAO, et al. Management of *Helicobacter pylori* infection—the Maastricht V/Florence consensus report. Gut. 2017;**66**:6-30

[22] Chan FKL, To KF, Wu JCY, et al. Randomised trial of eradication of *Helicobacter pylori* before non-steroidal anti-inflammatory drug therapy to prevent peptic ulcers. Lancet. 1997;**350**(9083):975-979

[23] Kalia N, Bardhan KD, Reed MWR, et al. Mechanisms of *Helicobacter pylori*-induced rat gastric mucosal microcirculatory disturbances in vivo. Digestive Diseases and Sciences. 2000;**45**(4):763-772

[24] Smolka AJ, Schubert ML. *Helicobacter pylori*-induced changes in gastric acid secretion and upper gastrointestinal disease. Current Topics in Microbiology and Immunology. 2017;**400**:227-252

[25] Stathis A, Chini C, Bertoni F, et al. Long-term outcome following *Helicobacter pylori* eradication in a retrospective study of 105 patients with localized gastric marginal zone B-cell lymphoma of MALT type. Annals of Oncology. 2009;**20**(6):1086-1093

[26] Floch P, Mégraud F, Lehours P. *Helicobacter pylori* strains and gastric MALT lymphoma. Toxins. 2017;**9**(4):132

[27] Hasni S, Ippolito A, Illei G. *Helicobacter pylori* and autoimmune diseases. Oral Diseases. 2011;**17**(7):621-627

[28] Øverby A, Murayama SY, Michimae H, et al. Prevalence of gastric non-*Helicobacter pylori-Helicobacters* in Japanese patients with gastric disease. Digestion. 2017;**95**(1):61-66

[29] Morgner A, Lehn N, Andersen LP, et al. *Helicobacter heilmannii*–associated primary gastric low-grade MALT lymphoma: Complete remission after curing the infection. Gastroenterology. 2000;**118**(5):821-828

[30] Park J, Greenberg E, Parsonnet J, et al. Summary of IARC working group meeting on *Helicobacter pylori* eradication as a strategy for preventing gastric cancer. IARC Work Group Report. 2014;**8**:1-4

[31] Stein M, Rappuoli R, Covacci A. Tyrosine phosphorylation of the *Helicobacter pylori* CagA antigen after cag-driven host cell translocation. Proceedings of the National Academy of Sciences of the United States of America. 2000;**97**(3):1263-1268

[32] Stein M, Bagnoli F, Halenbeck R, et al. c-Src/Lyn kinases activate *Helicobacter pylori* CagA through tyrosine phosphorylation of the EPIYA motifs. Molecular Microbiology. 2002;**43**(4):971-980

[33] Cover TL, Peek RM Jr. Diet, microbial virulence, and *Helicobacter pylori*-induced gastric cancer. Gut Microbes. 2013;**4**(6):482-493

[34] Armstrong H, Bording-Jorgensen M, Dijk S, et al. The complex interplay between chronic inflammation, the microbiome, and cancer: Understanding disease progression and what we can do to prevent it. Cancers. 2018;**10**(3):83

[35] Malfertheiner P, Sipponen P, Naumann M, et al. *Helicobacter pylori* eradication has the potential to prevent gastric cancer: A state-of-the-art critique. The American Journal of Gastroenterology. 2005;**100**(9):2100-2115

[36] Neunert C, Lim W, Crowther M, et al. The american society of hematology 2011 evidence-based practice guideline for immune thrombocytopenia. Blood. 2011;**117**(16):4190-4207

[37] Takahashi T, Yujiri T, Inoue Y, et al. Molecular mimicry by *Helicobacter pylori* CagA protein may be involved in the pathogenesis of *H. pylori* -associated chronic idiopathic thrombocytopenic purpura. British Journal of Haematology. 2004;**124**:91-96

[38] Stasi R, Sarpatwari A, Segal JB, et al. Effects of eradication of *Helicobacter pylori* infection in patients with immune thrombocytopenic purpura: A systematic review. Blood. 2008;**113**:1231-1240

[39] Kim H, Lee W, Lee K, et al. Efficacy of *Helicobacter pylori* eradication for the 1st line treatment of immune thrombocytopenia patients with moderate thrombocytopenia. Annals of Hematology. 2015;**94**:739-746

[40] Shino H, Brito H, Aparecida J, et al. *Helicobacter pylori* infection and immune thrombocytopenic purpura in children and adolescents: A randomized controlled trial. Platelets. 2015;**26**(4):336-341

[41] Jones NL, Koletzko S, Goodman K, et al. Joint ESPGHAN/NASPGHAN guidelines for the management of *Helicobacter pylori* in children and adolescents. Journal of Pediatric Gastroenterology and Nutrition. 2017;**64**(6):991-1003

[42] Cardenas VM, Mulla ZD, Ortiz M, et al. Iron deficiency and *Helicobacter pylori* infection in the United States. American Journal of Epidemiology. 2018;**163**(2):127-134

[43] Dubois S, Kearney DJ. Iron-deficiency anemia and *Helicobacter pylori* infection: A review of the evidence. The American Journal of Gastroenterology. 2005;**100**:453-459

[44] Milman N, Rosenstock S, Andersen L, et al. Serum ferritin, hemoglobin, and *Helicobacter pylori* infection: A seroepidemiologic survey comprising 2794 Danish adults. Gastroenterology. 1998;**115**:268-274

[45] Papagiannakis P, Michalopoulos C, Papalexi F, et al. The role of *Helicobacter pylori* infection in hematological disorders. European Journal of Internal Medicine. 2018;**24**(8):685-690

[46] Qu X, Huang X, Xiong P, et al. Does *Helicobacter pylori* infection play a role in iron deficiency anemia? A meta-analysis. World Journal of Gastroenterology. 2010;**16**(7):886-896

[47] Hershko C, Ronson A, Souroujon M, et al. Variable hematologic presentation of autoimmune gastritis: Age-related progession from iron deficiency to cobalamin depletion. Blood. 2006;**107**:1673-1679

[48] Goddard AF, James MW, Mcintyre AS, et al. Guidelines for the management of iron deficiency anaemia. Gut. 2011;**60**:1309-1316

[49] Chey WD, Leontiadis GI, Howden CW, et al. CME ACG clinical guideline: Treatment of *Helicobacter pylori* infection. The American Journal of Gastroenterology. 2017;**112**(2): 212-239

[50] Stopeck A. Links between *Helicobacter pylori* infection, cobalamin deficiency, and pernicious anemia. Archives of Internal Medicine. 2000;**160**:1229-1230

[51] DeLuca VA. *Helicobacter pylori* gastric atrophy and pernicious anemia. Gastroenterology. 1992;**102**(2):744-745

[52] Kaptan K, Beyan C, Ural AU, et al. *Helicobacter pylori*—Is it a novel causative agent in vitamin B12 deficiency? Archives of Internal Medicine. 2000;**160**(9):1349

[53] Kucukazman M, Yeniova O, Dal K, et al. *Helicobacter pylori* and cardiovascular disease. European Review for Medical and Pharmacological Sciences. 2015;**19**(19):3731-3741

[54] Sharma V, Aggarwal A. *Helicobacter pylori*: Does it add to risk of coronary artery disease. World Journal of Cardiology. 2015;**7**(1):19

[55] Lee M, Baek H, Park JS, et al. Current *Helicobacter pylori* infection is significantly associated with subclinical coronary atherosclerosis in healthy subjects: A cross-sectional study. PLoS One. 2018;**13**(3):e0193646

[56] Jukic A, Bozic D, Kardum D, et al. *Helicobacter pylori* infection and severity of coronary atherosclerosis in patients with chronic coronary artery disease. Therapeutics and Clinical Risk Management. 2017;**13**:933-938

[57] Sagud M, Vlatkovic S, Strac DS, et al. Latent *Toxoplasma gondii* infection is associated with decreased serum triglyceride to high-density lipoprotein cholesterol ratio in male patients with schizophrenia. Comprehensive Psychiatric Care. 2018;**82**:115-120

[58] Kelesidis T, Oda MN, Borja MS, et al. Predictors of impaired HDL function in HIV-1 infected compared to uninfected individuals. The Journal of Acquired Immune Deficiency Syndromes. 2017;**75**(3):354-363

[59] Sayyahfar S, Davoodzadeh F, Hoseini R, et al. Comparison of tuberculin skin test and interferon gamma release assay for diagnosis of latent tuberculosis infection in pediatric candidates of renal transplantation. Pediatric Transplantation. 2018;**22**(2):e13148

[60] Bulajic M, Panic N, Löhr JM. *Helicobacter pylori* and pancreatic diseases. World Journal of Gastrointest Pathophysiology. 2014;**5**(4):380-383

[61] Raderer M, Wrba F, Kornek G, et al. Association between *Helicobacter pylori* infection and pancreatic cancer. Oncology. 1998;**55**(16):16-19

[62] Stolzenberg-Solomon RZ, Blaser MJ, Limburg PJ, et al. *Helicobacter pylori* seropositivity as a risk factor for pancreatic cancer. The Journal of the National Cancer Institute. 2001;**93**(12):937-941

[63] Rabelo-Gonçalves EM, Roesler BM, Zeitune JM. Extragastric manifestations of *Helicobacter pylori* infection: Possible role of bacterium in liver and pancreas diseases. World Journal of Hepatology. 2015;**7**(30):2968-2979

[64] Haarstad H, Petersen H. Short- and long-term effects of secretin and a cholecystokinin-like peptide on pancreatic growth and synthesis of RNA and polyamines. Scandinavian Journal of Gastroenterology. 1989;**24**(6):721-732

[65] Culver EL, Smit WL, Evans C, et al. No evidence to support a role for *Helicobacter pylori* infection and plasminogen binding protein in autoimmune pancreatitis and IgG4-related disease in a UK cohort. Pancreatology. 2017;**17**:395-402

[66] Ioannou GN, Weiss NS, Kearney DJ. Is *Helicobacter pylori* seropositivity related to body mass index in the United States? Alimentary Pharmacology and Therapeutics. 2005; **21**:765-772

[67] Osawa H, Nakazato M, Date Y, et al. Impaired production of gastric ghrelin in chronic gastritis associated with *Helicobacter pylori*. The Journal of Clinical Endocrinology and Metabolism. 2005;**90**(1):10-16

[68] Nwokolo CU, Freshwater DA, O'Hare P, et al. Plasma ghrelin following cure of *Helicobacter pylori*. Gut. 2003;**52**(5):637-640

[69] Sakata I, Sakai T. Ghrelin cells in the gastrointestinal tract. International Journal of Peptide. 2010;**2010**:1-7

[70] Tatsuguchi A, Miyake K, Gudis K, et al. Effect of *Helicobacter pylori* infection on ghrelin expression in human gastric mucosa. The American Journal of Gastroenterology. 2004;**99**(11):2121-2127

[71] Churm R, Davies J, Stephens J, et al. Ghrelin function in human obesity and type 2 diabetes: A concise review. Obesity Reviews. 2017;**18**(2):140-148

[72] Wong F, Rayner-Hartley E, Byrne MF. Extraintestinal manifestations of *Helicobacter pylori*: A concise review. World Journal of Gastroenterology. 2014;**20**(34):11950-11961

[73] He C, Yang Z, Lu N-H. *Helicobacter pylori* infection and diabetes: Is it a myth or fact? World Journal of Gastroenterology. 2014;**20**(16):4607-4617

[74] Loffeld RJLF. *Helicobacter pylori*, obesity and gastro-oesophageal reflux disease. Is there a relation? A personal view. The Netherlands Journal of Medicine. 2005;**63**(9):344-347

[75] Kountouras J, Boziki M, Gavalas E, et al. Eradication of *Helicobacter pylori* may be beneficial in the management of Alzheimer's disease. Journal of Neurology. 2009;**256**(5):758-767

[76] Malaguarnera M, Bella R, Alagona G, et al. *Helicobacter pylori* and Alzheimer's disease: A possible link. European Journal of Internal Medicine. 2004;**15**(6):381-386

[77] Roubaud-Baudron C, Krolak-Salmon P, Quadrio I, et al. Impact of chronic *Helicobacter pylori* infection on Alzheimer's disease: Preliminary results. Neurobiology of Aging. 2012;**33**:1009.e11-1009.e19

[78] Kountouras J, Boziki M, Zavos C, et al. A potential impact of chronic *Helicobacter pylori* infection on Alzheimer's disease pathobiology and course. Neurobiology of Aging. 2012; **33**:e3-e4

[79] Weller C, Charlett A, Oxlade NL, et al. Role of chronic infection and inflammation in the gastrointestinal tract in the etiology and pathogenesis of idiopathic parkinsonism. Part 3: predicted probability and gradients of severity of idiopathic parkinsonism based on *H. pylori* antibody profile. Helicobacter. 2005;**10**(4):288-297

[80] Tradtrantip L, Zhang H, Saadoun S, et al. Anti-aquaporin-4 monoclonal antibody blocker therapy for neuromyelitis optica. Annals of Neurology. 2012;**71**(3):314-322

[81] Chen Y, Blaser MJ. Inverse associations of *Helicobacter pylori* with asthma and allergy. Archives of Internal Medicine. 2007;**167**(8):821

[82] Pacifico L, Osborn JF, Tromba V, et al. *Helicobacter pylori* infection and extragastric disorders in children: A critical update. World Journal of Gastroenterology. 2014;**20**(6): 1379-1401

[83] Zhou X, Wu J, Zhang G. Association between *Helicobacter pylori* and asthma: A meta-analysis. European Journal of Gastroenterology and Hepatology. 2013;**25**(4):460-468

[84] D'Elios MM, Codolo G, Amedei A, et al. *Helicobacter pylori*, asthma and allergy. FEMS Immunology and Medical Microbiology. 2009;**56**(1):1-8

[85] Arnold IC, Dehzad N, Reuter S, et al. *Helicobacter pylori* infection prevents allergic asthma in mouse models through the induction of regulatory T cells. The Journal of Clinical Investigation. 2011;**121**(8):3088-3093

[86] Fox JG, Yan LL, Dewhirst FE, et al. *Helicobacter bilis* sp. nov., a novel *Helicobacter* species isolated from bile, livers, and intestines of aged, inbred mice. The Journal of Clinical Microbiology. 1995;**33**(2):445-454

[87] Fox JG, Yan L, Shames B, et al. Persistent hepatitis and enterocolitis in germfree mice infected with *Helicobacter hepaticus*. Infection and Immunity. 1996;**64**(9):3673-3681

[88] Franklin CL, Beckwith CS, Livingston RS, et al. Isolation of a novel *Helicobacter* species, *Helicobacter cholecystus* sp. nov., from the gallbladders of Syrian hamsters with cholangiofibrosis and centrilobular pancreatitis. Journal of Clinical Microbiology. 1996; **34**(12):2952-2958

[89] Al-Soud WA, Stenram U, Ljungh A, et al. DNA of *Helicobacter* spp. and common gut bacteria in primary liver carcinoma. Digestive and Liver Disease. 2008;**40**:126-131

[90] Kobayashi T, Harada K, Miwa K, et al. *Helicobacter* genus DNA fragments are commonly detectable in bile from patients with extrahepatic biliary diseases and associated with their pathogenesis. Digestive Diseases and Sciences. 2005;**50**(5):862-867

[91] Fukuda K, Kuroki T, Tajima Y, et al. Comparative analysis of *Helicobacter* DNAs and biliary pathology in patients with and without hepatobiliary cancer. Carcinogenesis. 2002;**23**(11):1927-1932

[92] Segura-López FK, Güitrón-Cantú A, Torres J. Association between *Helicobacter* spp. infections and hepatobiliary malignancies: A review. World Journal of Gastroenterology. 2015;**21**(5):1414-1423

[93] Amodio P, Montagnese S, Gatta A, et al. Characteristics of minimal hepatic encephalopathy. Metabolic Brain Disease. 2004;**19**(3/4):253-267

[94] Holzinger F, Z'graggen K, Büchler MW. Mechanisms of biliary carcinogenesis: A pathogenetic multi-stage cascade towards cholangiocarcinoma. Annals of Oncology. 1999; **10**:122-126

[95] Bassi V, Marino G, Iengo A, et al. Autoimmune thyroid diseases and *Helicobacter pylori*: The correlation is present only in Graves's disease. World Journal of Gastroenterology. 2012;**18**(10):1093-1097

[96] Bertalot G, Montresor G, Tampieri M, et al. Decrease in thyroid autoantibodies after eradication of *Helicobacter pylori* infection. Clinical Endocrinology. 2004;**61**(5):650-652

[97] Abenavoli L, Arena V, Giancotti F, et al. Celiac disease, primary biliary cirrhosis and *Helicobacter pylori* infection: One link for three diseases. International Journal of Immunopathology and Pharmacology. 2010;**23**(4):1261-1265

[98] Ponzetto A, Cardaropoli S, Piccoli E, et al. Pre-eclampsia is associated with *Helicobacter pylori* seropositivity in Italy. Journal of Hypertension. 2006;**24**(12):2445-2449

[99] Moretti E, Figura N, Collodel G, et al. Can *Helicobacter pylori* infection influence human reproduction? World Journal of Gastroenterology. 2014;**20**(19):5567-5574

[100] Bellos I, Daskalakis G, Pergialiotis V. *Helicobacter pylori* infection increases the risk of developing preeclampsia: A meta-analysis of observational studies. International Journal of Clinical Practice. 2018;**72**(2):e13064

[101] Shiadeh MN, Moghadam ZB, Adam I, et al. Human infectious diseases and risk of preeclampsia: An updated review of the literature. Infection. 2017;**45**(5):589-600

[102] Parte AC. LPSN—list of prokaryotic names with standing in nomenclature. Nucleic Acids Research. 2014;**42**(D1):D613-D616

[103] Zhang R-G et al. Role of Helicobacter pylori infection in pathogenesis of gastric carcinoma. The World Journal of Gastrointestinal Pathophysiology. 2016;**7**(1):97-107

[104] Álvarez-Arellano L, Maldonado-Bernal C. Helicobacter pylori and neurological diseases: Married by the laws of inflammation. The World Journal of Gastrointestinal Pathophysiology. 2014;**5**(4):400-404

[105] Chiela M, Conciarz W, et al. Molecular mimicry in Helicobacter pylori infections. World Journal of Gastroenterology. 2017;**23**:3964-3977

Chapter 3

Endoscopical Aspects of *Helicobacter pylori* Gastritis in Children

Felicia Galoș, Cătălin Boboc, Gabriela Năstase,
Anca Orzan, Cristina Coldea, Mălina Anghel and
Mihaela Bălgrădean

Additional information is available at the end of the chapter

http://dx.doi.org/10.5772/intechopen.81437

Abstract

The current gold standard for the detection of *Helicobacter pylori* in children remains upper endoscopy plus mucosal biopsies. Endoscopy has the advantage of being able to detect complications of *Helicobacter pylori* infection and to rule out other upper gastro-intestinal pathologies. An additional advantage of endoscopy with gastric biopsy is that it allows physicians to obtain mucosa for urease testing, histological examination and bacterial culture. In children, there is a high correlation between antral nodularity at endoscopy and the presence of *Helicobacter pylori* infection. The authors have proposed to investigate the correlations between macroscopic aspects during endoscopy and histological findings, in order to identify those endoscopic and histopathological features that can help the clinician in clinical practice.

Keywords: *Helicobacter pylori*, children, gastritis, endoscopic aspects

1. Introduction

Helicobacter pylori (*H. pylori*) infection is acquired in childhood and remains an important cause of peptic ulcer disease (PUD) and gastric cancer. In comparison with adults, children and adolescents, however, infrequently develop these complications [1].

It is now well accepted that PUD, the most common stomach disease, is an infectious disease, and all consensus conferences agree that the causative agent, *H. pylori*, must be treated whit antibiotics [2].

The public health importance of *H. pylori* discovery, in 1982, and its role in stomach disease was recognized in 2005 with the attribution of the Nobel Prize in Medicine to Barry Marshall and Robin Warren. *H. pylori* was classified as a class I human carcinogen by World Health Organization in 1994.

Numerous diagnostic tests are available for detecting *H. pylori* infection: invasive techniques, which means endoscopy with biopsies for a rapid urease test (RUT), histology, culture and non-invasive techniques, such as serology, ^{13}C-Urea breath test (^{13}C-UBT), and the stool antigen test. There is no single method to detect *H. pylori* infection reliably and accurately. The choice of the diagnostic method depends on patients' age and complaints, technical difficulty level, costs and extensive accessibility in hospitals.

The same diagnostic methods used for adults can be used for children. However, *H. pylori* infection has certain particularities in children which have implications for diagnostic testing. *H. pylori* infection may slowly establish itself, so it is possible, in rare instances, to find the bacteria without traces of inflammation. At endoscopy, antral nodularity is common [2]. Histology provides an excellent diagnostic accuracy, allowing for the detection of the bacteria as well as for the grading of gastritis. The sensitivity and specificity of histology for the diagnosis depends on clinical settings, density of colonization and on pathologist's experience [3].

2. Material and methods

2.1. Patients

This was a prospective, single center study (in Maria Sklodowska Curie Children's Emergency Hospital Bucharest, Romania) that evaluated consecutive children referred by their physicians for upper endoscopy because of dyspepsia. They were all screened for *H. pylori* and had a positive stool antigen test.

Demographic characteristics and family history of each patient were collected through a questionnaire, which was completed by parents or by patients depending on the age of the child. Demographic data included patients' age, gender, and residency (urban or country area). Information on patient's history of *H. pylori* infection as well as on previous therapies was obtained. History of siblings or parents infection was also assessed. Patients were asked about the time of onset and duration of gastrointestinal symptoms, use of proton pump inhibitors, H_2 receptors antagonists, non steroidal anti-inflammatory drugs or steroidal drugs. Smoking status and alcohol consumption was determined as well.

Excluding criteria were: use of proton pump inhibitors or H_2 receptors antagonists and antibiotics as well as non steroidal anti inflammatory drugs or steroidal therapy 2 weeks before the beginning of the study, history of intestinal surgery (except for polypectomy and appendectomy), concomitant severe disease (heart, lungs, kidneys and endocrine diseases), and smoking and alcohol consumption.

The study was approved by Ethics Committee.

2.2. Endoscopy

All patients underwent endoscopy with biopsy specimens for histology (one for the antrum, one for the corpus). One sample from the antrum was used for rapid urease test. Two additional biopsies were taken from the antrum for bacterial culture. The samples were placed in separates vials, previously identified, containing the appropriate medium for each test.

This procedure was performed in patients with a minimum of 10 hours of fasting, under general anesthesia or conscious sedation. Vital signs were continuously monitored for the entire procedure.

Written informed consent was obtained from the parent or tutor of each child included in the study.

2.3. Histology

A biopsy of gastric body and antrum were fixed in a solution of formaldehyde 10%. Subsequently, the gastric mucosa samples were processed, following the usual steps of dehydration and paraffin embedding.

Two stains were used for histological study: hematoxylin eosin and Giemsa. Hematoxylin eosin stain was used to evaluate inflammatory cells and *H. pylori*. Giemsa stain was needed when hematoxylin eosin stain failed to identify the bacterium. The Giemsa stain is the preferred stain for detecting *H. pylori* because of its technical simplicity, high sensitivity and low cost.

Gastritis was graded according to the Sydney System [6] that assesses the severity of inflammation, the level of activity (the degree of polymorph neutrophil inflammation), and the presence of atrophy and of intestinal metaplasia on a scale from 0 to 3.

In accordance with the Sydney System, the density of *H. pylori* infection was also semi quantitatively classified on a scale from 0 to 3 (mild, moderate, and marked).

H. pylori was recognized in the histological section appearing as a short curved or spiral bacillus resting on the epithelial surface or in the mucus layer.

2.4. Bacterial culture

The biopsy specimens collected for bacterial culture were transported in commercial selective transport *H. pylori* medium, Portagerm pylori (BioMérieux SA, Marcy l'Etoile, France), and were inoculated after a few hours onto selective medium Pylori Agar (BioMérieux Italia). The plates were incubated under microaerobic condition at 37° for 72 h. Once incubated, the colonies resembling *H. pylori* were identified by Gram stain and by oxidase, catalase and urease tests. Suspensions from the primary plates were prepared in sterile solution to perform an E-test on Pylori Agar. An agar plate was streaked in three directions with a swab dipped into each bacterial suspension to produce a lawn of growth, an E-Test strip (E-Test; AB Bio disk, Solna, Sweden) was placed each onto separate plate, which was immediately incubated in a microaerobic atmosphere at 37°C for 72 h. Isolated strains were tested for amoxicillin, clarithromycin, metronidazole, and levofloxacin resistance following the recommendations of the European Committee on Antimicrobial Susceptibility Testing.

2.5. Statistical analysis

The data was collected and analyzed with Microsoft Excel 2013 and PSPP version 1.0.1. Continuous variables with a normal distribution were expressed as a mean with standard derivation (SD) and continuous variables with a non-normal distribution as median with interquartile range (IQR). Differences between groups were analyzed using Student t-test and Mann-Whitney U test for continuous variables, and Fisher's exact test for categorical variables. A p value <0.05 was considered statistically significant for all the analyzed parameters.

3. Results

Of the 38 patients who underwent upper endoscopy with biopsies by protocol (**Figure 1**), nine were excluded because of negative results in both culture and histology.

In the study, the culture and histology examination findings were accepted as "gold standard". The detection of *H. pylori* in at least one of the two tests was accepted as *H. pylori* positivity. Negative results in both culture and histology were accepted as *H. pylori* negativity.

Twenty-nine cases (76.31%) were included in the final analyses, 19 females (65.51%) and the 10 males (34.49%). The ages were between 3 years and 7 months and 17 years and 8 months (mean age 13.5 ± 4.53 years).

Four patients had a family history of peptic ulcer disease. In 15 children the duration of symptoms was more than 6 months and 12 patients were previously treated for *H. pylori* (**Table 1**).

The mean duration of the period between the onset of symptoms and the effective diagnosis in patients with a family history of upper gastrointestinal diseases was 3.75 ± 3.69 and 8.66 ± 5.42 months in those with negative family history (p = 0.17). A family history of gastric or duodenal ulcer did not significantly alter the length of time between the onset of symptoms and the diagnosis according to our statistical results, which, however, may have been influenced by the restricted number of patients in our study population.

Twelve patients had previous therapies. The median age of patients who were previously treated was 14.5 ± 3.74 and 13 ± 4.71 years old of those without any anterior therapy (p = 0.2).

The most common finding identified at endoscopy was macroscopic nodular antral gastritis, which was present in 22 patients (75.86%) (**Figure 2**). Among these, 10 had additional associated macroscopic lesions: 8 presented with nodular gastritis of gastric body, 1 with bulbitis, and one with esophagitis. Endoscopy showed antral hyperemia in 4 cases and a normal mucosal aspect in other 3 cases (**Table 2**).

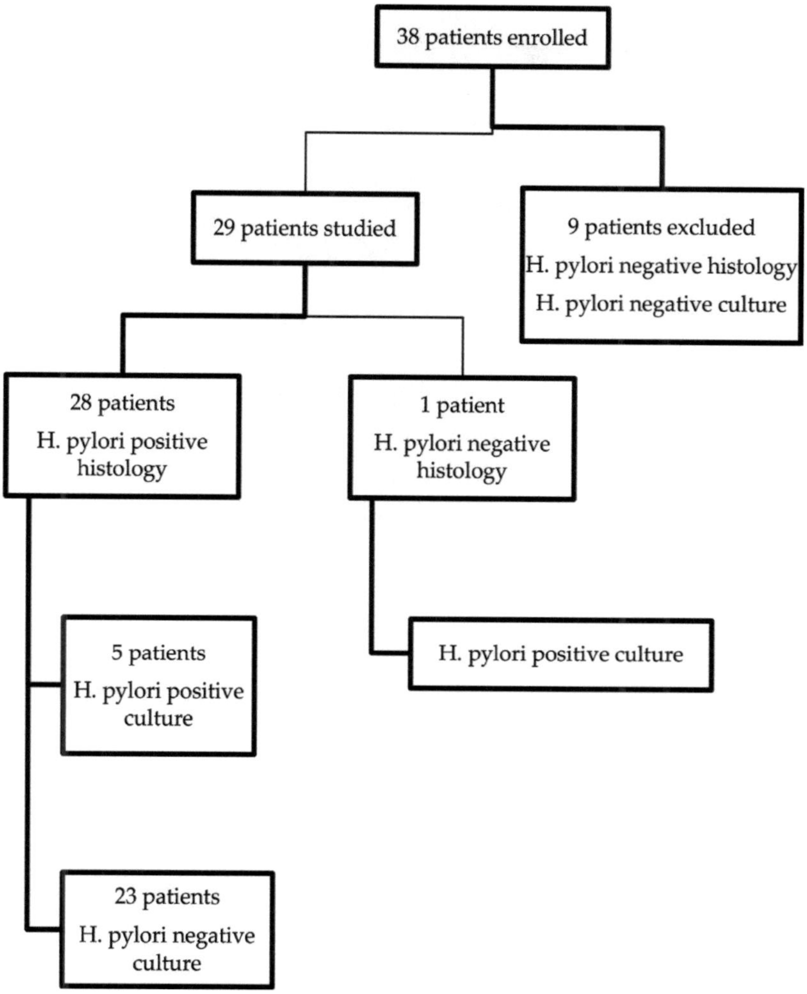

Figure 1. Flow chart of the study.

Mean age ± SD, years	13.5 ± 4.47
Male/female	10/29
Familial history for *H. pylori* infection	4/29
Peptic ulcer/non ulcer dyspepsia	1/28
Previous therapy	12/29

Table 1. Clinical and demographical characteristics of the patients.

We tried to find out if there was a significant difference in the severity of endoscopic findings between patients who received previous therapy and those who did not. Among the 12 previously treated patients, 7 (58.33%) presented with macroscopic nodular antral gastritis,

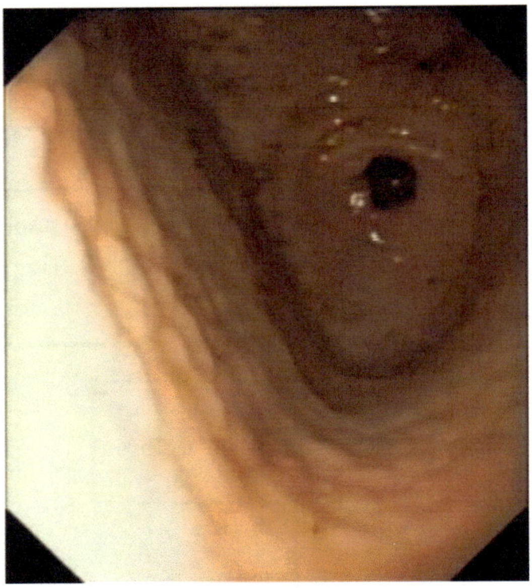

Figure 2. Endoscopical aspect of *Helicobacter pylori* gastritis in children (macroscopic nodular antral gastritis).

Endoscopic features	n (%)
Macroscopic nodular antral gastritis	22 (75.86%)
Nodular antral gastritis (only)	12
Nodular gastritis of corpus (with)	8
Erosive bulbitis (with)	1
Esophagitis (with)	1
Antral hyperemia without macroscopic nodularity	4 (13.79%)
Normal	3 (10.34)
Total	29 (100%)

Table 2. Endoscopic features associated to *Helicobacter pylori* infection.

2 (16.67%) with antral hyperemia and 3 (25%) showed a normal mucosal aspect. In patients who were not previously treated, we observed macroscopic nodular antral gastritis in 15 cases (88.24%), antral hyperemia in 2 cases (11.76%) while a normal appearance of gastric mucosa was never detected (**Table 3**). There was not a statistically significant association between the severity of mucosal damage at endoscopy and the existence of a previous therapy against the infection ($p = 0.06$).

In our study, bleeding was the presenting symptom in 4 children; three of them had pan gastritis, and one had nodular gastritis and esophagitis.

http://dx.doi.org/10.5772/intechopen.81437

Endoscopic features	*H. pylori* infection without anterior therapies, n (%)	*H. pylori* infection with anterior therapies, n (%)
Macroscopic nodular antral gastritis	15 (88.24)	7 (58.33%)
Nodular antral gastritis (only)	8	4
Nodular gastritis of corpus (with)	6	2
Erosive bulbitis (with)	0	1
Esophagitis (with)	1	2 (16.67%)
Antral hyperemia without macroscopic nodularity	2 (11.76)	3 (25.0%)
Normal	0	12 (100%)
Total	17 (100)	

Table 3. Endoscopic features associated to *Helicobacter pylori* infection: without anterior therapies versus with anterior therapies.

4. Discussion

There is a clear association between *H. pylori* and gastritis, gastric ulcer, and duodenal ulcers. Studies have shown that this pathogen causes mucosa-associated lymphoid tissue (MALT) lymphoma in both children and adults. In fact, when the organism is eradicated, extra gastric metastases or sites of MALT lymphoma resolve [4].

The finding of *H. pylori*-associated gastritis without duodenal or gastric mucosal lesions puts the pediatric gastroenterologist in a dilemma on recommending eradication treatment. *H. pylori*-associated gastritis without PUD rarely gives rise of symptoms or progresses to severe complications of the disease during childhood [5]. The risk of *H. pylori*- associated cancer or MALT-lymphoma during childhood is extremely low in Europe and North America. Ohno reported two cases in Japanese children, a 14-year-old boy and another 6-year-old boy with MALT and *H. pylori* infection [5]. The lower risk of complications in children may be in part explained by a different immune response against the infection. In comparison with adults, gastric biopsies obtained from children infected with *H. pylori* show a lower degree of inflammation. In addition, a higher number of immunosuppressive regulatory T cells and a more prominent IL-10 mediated anti-inflammatory response have been detected in pediatric patients [1].

During childhood, *H. pylori* is associated with antral predominant gastritis and duodenal ulcers [6].

In our study, the most frequent lesion identified by endoscopy was macroscopic antral nodular gastritis, which was present in 22 patients (78.86%). This high frequency is in accordance with a retrospective study from Japan that also found out a marked prevalence of nodular antral gastritis associated with *H. pylori* infection (98.5%) [7]. A nodular antral gastritis frequency of 82.53% was also reported by a Turkish study conducted in adults and adolescents

[8]. In a pediatric polish study, the sensitivity of antral nodularity associated with *H. pylori* was 91.6%, and the specificity was 91% [9]. A slightly lower value of specificity of antral nodularity, similar to our results (75.86%), was detected by several authors and ranged from 64 to 85.2%. Higher specificity was found by others [7–9].

Although the mechanisms underlying nodular gastritis in children is not clear yet, it is thought that lymphoid follicles with germinal center form nodules on gastric mucosa or that inflammatory reaction generated by *H. pylori* infection results in an exaggerated appearance of a normal gastric mucosa [10].

In a 14-year-old boy we observed erosions at endoscopy. The frequency of these lesions in our study (3.45%) is similar to the one measured by another study conducted in Italy (3.40%) [11].

A prospective study, carried out during 1-month simultaneously in 19 centers among 14 European countries, showed a frequency of 8.1% of ulcers and/or erosions in children, occurring mainly in the second decade of life, but *H. pylori* infection and toxic gastric medications were less frequently implicated than expected in their development. On a total of 56 children with ulcers or erosions, *H. pylori* was present in 15 patients (27%), 8 used NSAIDs, 5 were treated with steroids, 5 with immune-suppressive drugs, 6 with antibiotics, 1 with antacids, 6 with H_2 blockers and 8 with proton pump inhibitors (more than one risk factor was detected in 32 of 56 children) [12].

For years, reports have noted an association between peptic ulcer disease and families with a strong history of upper gastrointestinal tract disease, in particular between gastric and duodenal ulcers. Family history of gastric cancer is an important component in the diagnosis and management of *H. pylori* infection in children. Children with a mother or a father with gastric cancer are considered to be at very high risk owing to shared genetic characteristics, environmental factors, and virulence features of the infecting strain of *H. pylori* [4].

In countries with an elevated risk for gastric cancer, however, eradicating *H. pylori* in childhood could be more effective in preventing gastric atrophy, and ultimately, cancer development [13]. It still remains to be determined whether *H. pylori*-infected children with gastric atrophy are at increased risk for gastric cancer [14].

Recently, a decreasing proportion of *H. pylori*-positive peptic ulcers in adults has been observed, along with a decrease in the prevalence of infection, while, on the other hand, an increasing number of patients that use non-steroidal anti-inflammatory drugs (NSAIDs) has been noted [7, 15]. Regarding children, there are a few available data in the literature that investigate the trend of *H. pylori* prevalence in peptic ulcer [16].

In our study, four patients had a family history of *H. pylori* infection, none of gastric cancer, two of peptic ulcer, and two of *H. pylori* chronic gastritis. In this situation, the period of the onset of symptoms and to presentation to the doctor was less than 3 months, so that the average duration of symptoms are 3.75 ± 3.69 months in comparison with 8.66 ± 5.42 months for those without family history for gastric or duodenal ulcers. Influence of family history for upper gastrointestinal tract diseases to the period of the onset of symptoms and diagnosis, can be explained by the consciousness of the disease and the risks than derive from it.

The sex difference between the *H. pylori*-positive and *H. pylori*-negative group is also of great interest. We found female preponderance in the study group (65.51%), similarly with another report in our geographic area (78.49%) [17].

Studies have unanimously shown a male preponderance for peptic ulcer disease in children. It is still not known why primary peptic ulcers predominantly develop in infected male children. Epidemiological studies do not suggest any sex predilection in *H. pylori* infection [16].

Median age for patients with previous therapies was 14.5 ± 3.74 years, comparative with 13 ± 4.71 years for patients without previous therapies, results or else expected. We do not have data to express if it is failure of antimicrobial therapy or reinfection. 1/12 patient with previous therapies had family history of peptic ulcer disease. We do not investigate all the family member of each child, and therefore we do not know the real status of *H. pylori* infection. Familial history for gastrointestinal disease was collected for interview. Magistà et al. [18] identified two variables by logistic regression analysis as predictors of *H. pylori* reinfection: age of primary infection and having an infected sibling. Multivariable analysis revealed that only age at primary infection correlates with an increased risk of reinfection [18].

In patients with anterior therapies, the endoscopic features were less serious than in those without any previous treatment. All three patients with normal endoscopic mucosa were anteriorly treated. These results suggest that children might become "tolerant" to the bacterium or that the growing child is more resistant to *H. pylori*-induced lesions. The evidence that *H. pylori* infection in children coexist with normal gastric mucosa was reported in a percent comparable with our results (11%) [19]. This is the reason for which we strongly recommend to take biopsies at least for histological exam in children and adolescents, even if a normal appearance of mucosa is observed during endoscopy.

The ability of *H. pylori* to manipulate the immune response (activation or inactivation of Toll-like receptors dependent response) may be responsible for bacterial survival and a mild course of infection in children [20].

5. Conclusions

The main endoscopic feature found in our study was macroscopic nodular antral gastritis, in 75.86%. In 10.34% of cases the endoscopic aspect of mucosa was normal. All patients with normal endoscopic mucosa were previously treated. These results suggest that children might become "tolerant" to the bacterium or that the growing child is more resistant to *H. pylori*-induced lesions.

Acknowledgements

The authors thank Dr. Augustina Enculescu for her histological support.

Conflict of interest

None declared.

Author details

Felicia Galoş[1,2*†], Cătălin Boboc[2,†], Gabriela Năstase[2], Anca Orzan[1,2], Cristina Coldea[1,2], Mălina Anghel[2] and Mihaela Bălgrădean[1,2]

*Address all correspondence to: felicia_galos@yahoo.com

1 University of Medicine and Pharmacy Carol Davila, Bucharest, Romania

2 Maria Sklodowska Curie Children's Emergency Hospital, Bucharest, Romania

† These two authors contributed equally to this work.

References

[1] Jones NL, Koletzko S, Goodman K, Bontems P, Cadranel S, Casswall T, et al. Joint ESPGHAN/NASPGHAN guidelines for the management of *H. pylori* infection in children and adolescents (update 2016). JPGN. 2017;**64**:991-1003. DOI: 10.1097/MPG.0000000000001594

[2] Mégraud F, Lehours P. *Helicobacter pylori* detection and antimicrobial susceptibility testing. Clinical Microbiology Reviews. 2007;**20**:280-322. DOI: 10.1128/CMR. 00033-06

[3] Ricci C, Holton J, Vaira D. Diagnosis of *Helicobacter pylori*: Invasive and non-invasive tests. Best Practice & Research. Clinical Gastroenterology. 2007;**21**:299-313. DOI: 10.1016/j.bpg.2006.11.002

[4] Gold BD, Gilger MA, Czinn S. New diagnostic strategies for detection of *Helicobacter pylori* infection in paediatric patients. Gastroenterology & Hepatology. 2014;**10**(12 Suppl 7): 1-18. DOI: 05US14EBP1368

[5] Sierra MS, Hastings EV, Goodman KJ. What do we know about benefits of *H. pylori* treatment in childhood? Gut Microbes. 2013;**4**:549-567. DOI: 10.4161/gmic.27000

[6] Pacifico L, Anania C, Osborn JF, Feraro F, Chiesa C. Consequences of *Helicobacter pylori* infection in children. World Journal of Gastroenterology. 2010;**16**(41):5181-5194. DOI: 10.3748/wjg.v16.i41.5181

[7] Kato S, Nishino Y, Ozawa K, Konno M, Maisawa S, Toyoda S, et al. The prevalence of *Helicobacter pylori* in Japanese children with gastritis or peptic ulcer disease. Journal of Gastroenterology. 2004;**39**(8):734-738. DOI: 10.1007/s00535-004-1381-2

[8] Cosgun Y, Yildirim A, Yucel M, Karakoc AE, Koca G, Gonultas A, et al. Evaluation of invasive and noninvasive methods for the diagnosis of *Helicobacter pylori* infection. Asian Pacific Journal of Cancer Prevention. 2016;**12**:5265-5272. DOI: 10.22034/APJCP.2016.17.12.5265

[9] Łazowska-Przeorek I, Kotowska M, Banasiuk M, Karolewska-Bochenek K, Banaszkiewicz A, Gawronska A, et al. Value of antral nodularity for the diagnosis of *Helicobacter pylori* infection in children. Medical Science Monitor. 2015;**21**:1827-1830. DOI: 10.12659/MSM.893467

[10] Yang HR. Update on the diagnosis of *Helicobacter pylori* infection in children: What are the differences between adults and children? Pediatric Gastroenterology, Hepatology & Nutrition. 2016;**19**(2):96-103. DOI: 10.5223/pghn.2016.19.2.96

[11] Oderna G, Mura S, Valori A, Brustia R. Idiopathic peptic ulcers in children. JPGN. 2009;**48**(3):268-270

[12] Kalach N, Bontems P, Koletzko S, Mourad-Baars P, Shcherbakov P, Celinska-Cedro D, et al. Frequency and risk factors of gastric and duodenal ulcers or erosions in children: A prospective 1-month European multicenter study. European Journal of Gastroenterology & Hepatology. 2010;**22**(10):1174-1181. DOI: 10.1097/MEG.0b13e32833d36de

[13] Kato S, Kikuchi S, Nakajima S. When does gastric atrophy develop in Japanese children? Helicobacter. 2008;**13**(4):278-281. DOI: 10.1111/j.1523-5378.2008.00611.x.

[14] Kato S, Nakajima S, Nishino Y, Ozawa K, Minoura T, Konno M, et al. Association between gastric atrophy and *Helicobacter pylori* infection in Japanese children: A retrospective multicenter study. Digestive Diseases and Sciences. 2006;**51**(1):99-104. DOI: 10.1007/s10620-006-3091-5

[15] Arents N, Thijs JC, van Zwet AA, Kleibeuker JH. Does the declining prevalence of *Helicobacter pylori* unmask patients with idiopathic peptic ulcer disease? Trends over an 8 years period. European Journal of Gastroenterology & Hepatology. 2004;**16**:779-783. DOI: 10.1097/01.meg.0000108367.19243.73

[16] Tam YH, Lee KH, To KF, Chan KW, Cheung ST. *Helicobacter pylori* positive versus *Helicobacter pylori* negative idiopathic peptic ulcers in children with their long-term outcomes. JPGN. 2009;**48**(3):299-305

[17] Mărginean CO, Cotoi OS, Pitea AM, Mocanu S, Mărginean C. Assessment of the relationship between *Helicobacter pylori* infection, endoscopic appearance and histological chnges of the gastric mucosa in children with gastritis (a single center experience). Romanian Journal of Morphology and Embryology. 2013;**54**(3):709-715

[18] Magistà AM, Ierardi E, Castellaneta S, Miniello VL, Lionetti E, Francavilla A, et al. Helicobacter status and symptoms assessment two years after eradication in pediatric patients from a high prevalence area. JPGN. 2005;**40**:312-318

[19] Nardone G, Staibano S, Rocco A, Mezza E, Balzano T, Salvatore G, et al. Effect of *Helicobacter pylori* on gastric cell proliferation and genomic stability in a paediatric population of southern Italy. Digestive and Liver Disease. 2001;**33**:743-749. DOI: 10.1016/S1590-8658(01)80690-3

[20] Michalkiewicz J, Helmin-Basa A, Grzywa R, Czerwionka-Szaflarska M, Szaflarska-Poplawaska A, Mierzwa G, et al. Innate immunity components and cytokines in gastric mucosa in children with *Helicobacter pylori* infection. In: Mediators of Inflammation. Hindawi Publishing Corporation; 2015. DOI: 10.1155/2015/176726

Chapter 4

The Importance of *H. pylori* Infection in Liver Diseases

Tadeusz Wojciech Łapiński

Additional information is available at the end of the chapter

http://dx.doi.org/10.5772/intechopen.79969

Abstract

The chapter is a review of current knowledge on the impact of *H. pylori* infection on the clinical course of patients with various forms of liver damage. *H. pylori* infection is found in 50–90% of the world population. The bacteria not only mainly contribute to occurrence of gastric mucosa inflammation but also to gastric ulcer and cancer. *H. pylori* contains an active antioxidative system, which not only neutralizes free radicals but also synthesizes specific VacA toxin, which leads to destruction and apoptosis of the cells. A specific system of bacterial CagA genes has a special role in carcinogenesis. There is an increasing number of reports describing lesions in the circulatory system, pancreas, or the skin, connected with *H. pylori* infection. Liver colonization by *H. pylori* happens after transmission of the bacteria from the stomach, with blood, through the portal vein or directly through the bile ducts. The bacteria promote liver function deterioration in the course of toxic injury, autoimmune inflammation, chronic HBV, and HCV infection. Infections among people with liver cirrhosis are especially dangerous. In this group of patients, *H. pylori* infection may significantly worsen liver function, leading to hyperammonemia, increased portal pressure, and development of esophageal varices. Thus, testing for and treating this infection among patients with liver cirrhosis is especially important.

Keywords: *H. pylori*, hepatitis, hepatic liver cirrhosis, liver and biliary tract cancer

1. Introduction

Helicobacter pylori (*H. pylori*) is a microaerophile described for the first time by Marshal and Warren in 1982. This Gram-negative bacillus is resistant to the activity of gastric acid. Active, vegetative form of the bacteria is spiral, while the sporulation form is granular [1]. The bacterium contains stable DNA and a system of very effective DNA repair.

In developed countries, infection by these bacteria is detected in about 50% of the population, while in developing countries, the percentage may even reach 90% [2]. Still, however, there are no definitive pointers as to risk factors of *H. pylori* infection.

H. pylori produces high amounts of ureases, enzymes catalyzing urea decomposition to ammonia. This is especially important for the neutralization of hydrochloric acid in the stomach, which contributes to the growth of the bacteria. Bacteria demonstrate also the ability to pump out H^+ ions from cells. This leads to changes in the pH in the stomach, which in turn causes destruction of gastric mucosa.

H. pylori has an active antioxidizing system, which neutralizes free radicals. The system contains catalase, superoxide dismutase, and specific proteins MdaB and NapA.

Bacterial DNA contains genes encoding cytotoxin synthesis system. VacA toxin (vacuolating toxin) is encoded by a changing system of genes in 40–60% of bacterial strains. This underlies different toxic properties. In epithelial cells, VacA promotes fusion of several lysosomes and formation of large vacuoles, which changes the construction of cytoskeleton. VacA toxin induces apoptosis of epithelial cells and shows highly immunogenic properties.

Specific system of CagA genes encodes synthesis of CagA toxin, demonstrating properties reorganizing cytoskeleton and cell shape. Moreover, the toxin controls transcription and proliferation of the cell, as well as inflammatory reaction. CagA toxin plays a very important role in carcinogenesis in the stomach and other organs, when CagA-synthesizing *H. pylori* is detected [3, 4].

H. pylori is mainly present on the surface of epithelial cells of the mucosa, in the prepyloric part of the stomach. It has cilia allowing transport into intercellular spaces, and thanks to produced adhesins, it adheres to cell surface [1].

H. pylori infection influences local (in the gastric mucosa) and systemic increase in proinflammatory cytokines IL-1, -2, -4, -6, -8, -10, -17, interferon-β, and TNF-α [5]. This leads not only to development of local inflammatory reaction, but also potentiates generalized inflammatory reactions in the organism. *H. pylori* causes chronic atrophic gastritis, metaplasia, and dysplasia, leading to the development of gastric cancer. According to World Health Organization (WHO), the bacteria are a class I carcinogen. *H. pylori* may also potentiate extragastric organ disturbances, exacerbating the diseases of cardiovascular system or metabolic diseases, deteriorating normal function of the liver, especially in patients with cirrhosis [6].

2. Organ pathologies connected with *H. pylori* infection

H. pylori infection, especially in the case of strains producing CagA toxin, promotes development of coronary sclerosis and increases the probability of angina pectoris and cardiac infarct [7]. Effect of CagA toxin on promotion of sclerotic changes in coronary arteries leads to exacerbation of coronary disease, which increases mortality caused by circulatory failure in the group of patients infected with these bacteria and not subjected to eradication [8, 9].

H. pylori infection among patients with type 2 diabetes presents more seriously compared to patients without diabetes [10]. Moreover, impact of this infection onto the development of chronic pancreatitis has been reported, which indirectly affects liver function [11].

H. pylori infection is associated with many skin conditions. Higher incidence of chronic urticaria, acne rosacea, idiopathic thrombocytopatic purpura, psoriasis, atopic dermatitis, and some other dermatological conditions among patients infected with *H. pylori* has been demonstrated [12].

3. Liver injury

Liver colonization by *H. pylori* happens after transmission of the bacteria from the stomach, with blood, through the portal vein or directly through the bile ducts [13].

Experimental studies performed on mice and rats infected with *H. pylori* have shown the effect of this infection onto up triggered fibrosis and development of liver cirrhosis [14]. One of the first studies performed in patients with chronic liver injury pointed to the presence of *Helicobacter* genus bacteria in the liver tissue in 26% of patients [15]. Current studies in patients infected with HBV, HCV, and patients with chronic noninfectious liver conditions point to much higher incidence of *H. pylori* or bacterial DNA in the liver tissue. Infections are thought to occur in effect of disturbances in patients' immune functions [16]. Both experimental and clinical studies demonstrate unfavorable effect of *H. pylori* infection onto the course of liver injury, especially exacerbated fibrosis. One of the reasons for this is the influence of infection onto metabolic changes connected with carbohydrate turnover, synthesis of high-energy compounds (mainly ATP), and increased concentration of proinflammatory cytokines.

4. Chronic HBV and HCV infections

The frequency of *H. pylori* infection among patients with chronic hepatitis B is around 30–80% [17]. *H. pylori* infection is confirmed in 79% of patients with postinflammatory liver cirrhosis connected with HBV infection [18]. Favorable effect of *H. pylori* eradication on the course of the disease, including increased platelet count, has been demonstrated in the studies on patients chronically infected with HBV, with compensated liver cirrhosis and thrombocytopenia [19].

Among people chronically infected with HBV with primary liver cancer, *H. pylori* infection is found in 69% of patients. In the group of patients with primary liver cancer, but without HBV infection, *H. pylori* infection is much less frequent, as it is found in 33% of patients. These observations consistently point to unfavorable effect of *H. pylori* infection among HBV-infected patients with liver cirrhosis onto the risk of occurrence of primary liver cancer. Frequency of *H. pylori* infection among patients with chronic hepatitis B correlates with the incidence of hepatocellular cancer, both in men and women [20]. Among this type of patients, fast progression of inflammatory changes in the liver is observed, as well as intensified fibrosis, which promotes occurrence of primary neoplastic lesions [17, 21].

Among patients infected with *H. pylori* and HBV, liver function is impaired (prothrombin time is extended, and AST activity and bilirubin concentration increased), and esophageal varices, ascites and hyperammonemia with hepatic encephalopathy occur much more frequently [18].

Evaluation of *H. pylori* infection among patients with chronic HCV infection is difficult. Some authors state that the frequency of *H. pylori* infection among the patients in this group is about 38%. Around 45% of those result from CagA-synthesizing bacteria. No significant differences are found in the morphological picture of HCV-infected liver, between patients with or without *H. pylori* infection. There is also no correlation between *H. pylori* infection and IL 28B polymorphism [22]. However, many other authors present other observations. *H. pylori* infection may be present in even 70% of patients chronically infected with HCV [23]. Meta-analysis of 20 studies demonstrated higher incidence of *H. pylori* infection among HCV-positive patients, compared to persons without viral infection [17, 24]. Much higher fibrosis, loss of cellular proteins, and glycogen was found in morphological studies of the liver from HCV-positive patients, if those were coinfected with *H. pylori*, compared to those without coinfection [25].

5. Autoimmune diseases and liver steatosis

Studies performed among patients with chronic viral autoimmune and toxic hepatitis and coexisting *H. pylori* infection demonstrated improvement in liver function, including decreased ALT and AST activity, after effective eradication of bacteria [26].

H. pylori infection is found in 20–50% of patients with AIH. Among patients with PBC, *H. pylori* infection is found more frequently than among people from the control group (54 vs. 31% $p = 0.01$). However, the effect of this infection on the course of PBC has not been elucidated [27].

Infection by these bacteria worsens the course of underlying disease; however, pathogenesis is not completely clear [28, 29]. In the group of patients with PBC and AIH, *H. pylori* infection may lead to precipitous, unfavorable progression of the disease [30].

Reports on the effect of *H. pylori* infection on lipid turnover disturbances, leading to hypertriglyceridemia and hypercholesterolemia with concomitant decrease of HDL level, are published more and more frequently. This is especially important for the metabolism of hepatocytes and their steatosis, as well as in the process of liver fibrosis [31]. Many reports point to the fact that *H. pylori* infection hastens the development of NAFLD [32]. It has been demonstrated that *H. pylori* infection among patients with NAFL results in the development of NASH. Eradication of the bacteria significantly facilitates the treatment of liver steatosis [33]. Moreover, it has been found that *H. pylori* infection and steatosis constitute the risk of more frequent occurrence of cholecystolithiasis and choledocholithiasis [34].

6. Liver cirrhosis, hepatocellular carcinoma (HCC), and cholangiocarcinoma (CCC)

Inflammation of gastric mucosa is a frequent complication of liver cirrhosis. Usually, occurrence of chronic inflammation is observed, described as portal gastropathy. *H. pylori* infection in the group of patients with liver cirrhosis impacts exacerbation of inflammatory changes in the stomach, which in turn worsens liver function. This is especially dangerous among patients with advanced liver injury. Studies performed on this group of patients point to high significance of cytopathic effect of *H. pylori* onto hepatocytes [16, 35]. *H. pylori* infection affects increase of portal tension, which is one of the main etiologies of development of esophageal varices [6, 36]. In effect, correlation between the frequency of *H. pylori* infection and advancement of esophageal varices is observed [37].

Although in some studies more frequent *H. pylori* infection among patients with liver cirrhosis cannot be confirmed [38]; however, meta-analysis was performed that included mainly patients with alcoholic liver cirrhosis, which argues for much more frequent occurrence of these bacteria among such patients. *H. pylori* infection is much more frequent among patients with postinflammatory liver cirrhosis (connected with HBV or HCV infection) [37]. Incidence of *H. pylori* infection among patients with liver cirrhosis and concomitant HCV infection increases proportionally to progressing liver failure [39]. Moreover, it has been demonstrated that the highest percentage of people infected with *H. pylori* among those with HCV infection is observed in the case of patients in whom HCC developed [17, 24]. Many pieces of information argue that concomitant infection with *H. pylori* and HCV increases the incidence of HCC. Eradication of these bacteria in patients with cirrhotic liver leads to the increase of platelet count and improves efficacy of antiviral therapy [23, 40]. In the current setting, when direct-acting antivirals are commonly used, this is probably not so important; however, such studies have not been performed.

H. pylori catalyzes the reaction of urea decomposition to ammonia and carbon dioxide; however, among patients with subclinical hepatic encephalopathy, infection with these bacteria does not change the concentration of ammonia in the blood [41]. These observations are inconsistent, because among patients with liver cirrhosis, especially postinflammatory cirrhosis, more frequent occurrence of symptomatic hepatic encephalopathy with hyperammonemia is observed in the case of patients infected with *H. pylori*, compared to patients without this infection [37, 42]. In the studies performed in patients with liver cirrhosis, a correlation between increasing ammonia blood concentration and *H. pylori* infection has been demonstrated. Moreover, ammonia blood concentration was higher among patients with liver cirrhosis infected with *H. pylori*, compared to patients not infected with these bacteria [37].

In experimental studies performed on dogs, an association between *H. pylori* infection and occurrence of hepatocellular carcinoma (HCC) has been evidenced [43]. Evaluation of the effect of *H. pylori* infection on liver carcinogenesis in humans shows that in 58% of patients with HCC and in 62% of patients with CCC in the liver tissue surrounding focal lesion, DNA of these bacteria can be detected [44]. *H. pylori* may disturb the balance between hepatocyte proliferation and activity of apoptosis in the liver. In effect, there is a higher risk of occurrence of neoplastic cells in the liver [45].

In the pathogenesis of biliary duct carcinoma, *H. pylori* infection affects proliferation of biliary duct epithelium and development of inflammatory reaction in these cells. Activation of reactive oxygen species (oxidative stress) and reactive nitrogen species, mainly 8-nitroguanine, in the cells is detected. These reactions damage DNA of stem cells, playing a key role in carcinogenesis [46]. A special role in the development of bile duct carcinoma is attributed to *H. pylori* producing CagA toxin [47].

7. Conclusions

H. pylori infection is detected significantly more often among patients with chronic liver injury. This is especially dangerous in patients with liver cirrhosis. In this group of patients, *H. pylori* infection may significantly worsen liver function, affecting hyperammonemia, increase in portal pressure, and development of esophageal varices. Testing for and treating this infection is of paramount importance for these patients.

Current research on the impact of *H. pylori* infection in patients with chronic liver damage is inadequate. This points to the desirability of further research, particularly among patients with severe liver damage.

Acknowledgements

Studies have been done, and the manuscript was written without the support of other persons and sponsors.

Conflict of interest

No conflict of interest to be declared.

Author details

Tadeusz Wojciech Łapiński

Address all correspondence to: twlapinski@wp.pl

Department of Infectious Diseases and Hepatology, Medical University of Bialystok, Bialystok, Poland

References

[1] Bachir M, Allem R, Tifrit A, Medjekane M, Drici AE, Diaf M, Douidi KT. Primary antibiotic resistance and its relationship with cagA and vacA genes in *Helicobacter pylori* isolates from Algerian patients. Brazilian Journal of Microbiology. 2018; **49**(3): 544-551. doi: 10.1016/j.bjm.2017.11.003

[2] Mitchell H, Katelaris P. Epidemiology, clinical impacts and current clinical management of *Helicobacter pylori* infection. The Medical Journal of Australia. 2016;**204**(10):376-380

[3] Hatakeyama M. Structure and function of *Helicobacter pylori* CagA, the first-identified bacterial protein involved in human cancer. Proceedings of the Japan Academy. Series B, Physical and Biological Sciences. 2017;**93**(4):196-219. DOI: 10.2183/pjab.93.013

[4] Wroblewski LE, Peek RM Jr. *Helicobacter pylori*, cancer, and the gastric microbiota. Advances in Experimental Medicine and Biology. 2016;**908**:393-408. DOI: 10.1007/978-3-319-41388-4_19

[5] Sun X, Xu Y, Wang L, Zhang F, Zhang J, Fu X, Jing T, Han J. Association between TNFA gene polymorphisms and *Helicobacter pylori* infection: A meta-analysis. PLoS One. 2016;**11**(1):e0147410. DOI: 10.1371/journal.pone.0147410

[6] Waluga M, Kukla M, Żorniak M, Bacik A, Kotulski R. From the stomach to other organs: *Helicobacter pylori* and the liver. World Journal of Hepatology. 2015;**7**(18):2136-2146. DOI: 10.4254/wjh.v7.i18.2136

[7] Suzuki H, Franceschi F, Nishizawa T, Gasbarrini A. Extragastric manifestations of *Helicobacter pylori* infection. Helicobacter. 2011;**16**(Suppl 1):65-69. DOI: 10.1111/j.1523-5378.2011.00883.x

[8] Figura N, Palazzuoli A, Vaira D, Campagna M, Moretti E, Iacoponi F, Giordano N, Clemente S, Nuti R, Ponzetto A. Cross-sectional study: CagA-positive *Helicobacter pylori* infection, acute coronary artery disease and systemic levels of B-type natriuretic peptide. Journal of Clinical Pathology. 2014;**67**(3):251-257. DOI: 10.1136/jclinpath-2013-201743

[9] Wang JW, Tseng KL, Hsu CN, Liang CM, Tai WC, Ku MK, Hung TH, Yuan LT, Nguang SH, Yang SC, Wu CK, Chiu CH, Tsai KL, Chang MW, Huang CF, Hsu PI, Wu DC, Chuah SK. Association between *Helicobacter pylori* eradication and the risk of coronary heart diseases. PLoS One. 2018;**13**(1):e0190219. DOI: 10.1371/journal.pone.0190219

[10] Yang YJ, Wu CT, Ou HY, Lin CH, Cheng HC, Chang WL, Chen WY, Yang HB, Lu CC, Sheu BS. Male non-insulin users with type 2 diabetes mellitus are predisposed to gastric corpus-predominant inflammation after *H. pylori* infection. Journal of Biomedical Science. 2017;**24**(1):82. DOI: 10.1186/s12929-017-0389-x

[11] Rabelo-Gonçalves EM, Roesler BM, Zeitune JM. Extragastric manifestations of *Helicobacter pylori* infection: Possible role of bacterium in liver and pancreas diseases. World Journal of Hepatology. 2015;**7**(30):2968-2979. DOI: 10.4254/wjh.v7.i30.2968

[12] Guarneri C, Lotti J, Fioranelli M, Roccia MG, Lotti T, Guarneri F. Possible role of *Helicobacter pylori* in diseases of dermatological interest. Journal of Biological Regulators and Homeostatic Agents. 2017;**31**(Suppl. 2):57-77

[13] Pellicano R, Menard A, Rizzetto M, Megraud F. *Helicobacter* species and liver diseases: Association or causation? The Lancet Infectious Diseases. 2008;**8**:254-260. DOI: 10.1016/S1473-3099(08)70066-5

[14] Goo MJ, Ki MR, Lee HR, Yang HJ, Yuan DW, Hong IH, Park JK, Hong KS, Han JY, Hwang OK, Kim DH, Do SH, Cohn RD, Jeong KS. *Helicobacter pylori* promotes hepatic fibrosis in the animal model. Laboratory Investigation. 2009;**89**:1291-1303. DOI: 10.1038/labinvest.2009.90

[15] Stalke P, Al-Soud WA, Bielawski KP, Bakowska A, Trocha H, Stepinski J, Wadstrom T. Detection of *Helicobacter* species in liver and stomach tissues of patients with chronic liver diseases using polymerase chain reaction-denaturing gradient gel electrophoresis

and immunohistochemistry. Scandinavian Journal of Gastroenterology. 2005;**40**:1032-1041. DOI: 10.1080/00365520510023251

[16] Silva LD, Rocha AM, Rocha GA, de Moura SB, Rocha MM, Dani R, de Melo FF, Guerra JB, de Castro LP, Mendes GS, Ferrari TC, Lima AS, Queiroz DM. The presence of *Helicobacter pylori* in the liver depends on the Th1, Th17 and Treg cytokine profile of the patient. Memórias do Instituto Oswaldo Cruz. 2011;**106**:748-754

[17] Wang J, Chen RC, Zheng YX, Zhao SS, Li N, Zhou RR, Huang Y, Huang ZB, Fan XG. *Helicobacter pylori* infection may increase the risk of progression of chronic hepatitis B disease among the Chinese population: A meta-analysis. International Journal of Infectious Diseases. 2016;**50**:30-37. DOI: 10.1016/j.ijid.2016.07.014

[18] Huang J, Cui J. Evaluation of *Helicobacter pylori* infection in patients with chronic hepatic disease. Chinese Medical Journal. 2017;**130**(2):149-154. DOI: 10.4103/0366-6999.197980

[19] Zhang XH, He Y, Feng R, Xu LP, Jiang Q, Jiang H, Lu J, Fu HX, Liu H, Wang JW, Wang QM, Feng FE, Zhu XL, Xu LL, Xie YD, Ma H, Wang H, Liu KY, Huang XJ. *Helicobacter pylori* infection influences the severity of thrombocytopenia and its treatment response in chronic hepatitis B patients with compensatory cirrhosis: A multicenter, observational study. Platelets. 2016;**27**(3):223-229. DOI: 10.3109/09537104.2015.1077946

[20] Wang L, Zollinger T, Zhang J. Association between *Helicobacter pylori* infection and liver cancer mortality in 67 rural Chinese counties. Cancer Causes & Control. 2013;**24**(7): 1331-1337. DOI: 10.1007/s10552-013-0211-3

[21] Mohamed AA, Elshimy AA, El Sadik AO, Ezzat E, Nasar M, Elshaer SSM, Sayed MM. Association between severity of liver disease, frequency of *Helicobacter pylori* infection, and degree of gastric lesion in Egyptian patients with hepatitis B virus infection. The American Journal of Tropical Medicine and Hygiene. 2018;**98**(1):221-226. DOI: 10.4269/ajtmh.17-0291

[22] Gutwerk A, Wex T, Stein K, Langner C, Canbay A, Malfertheiner P, Link A. *Helicobacter Pylori* serology in relation to hepatitis C virus infection and IL28B single nucleotide polymorphism. Journal of Clinical Medicine. 2018;**7**(3):44-55. doi:10.3390/jcm7030044

[23] Hanafy AS, El Hawary AT, Hamed EF, Hassaneen AM. Impact of *Helicobacter pylori* eradication on refractory thrombocytopenia in patients with chronic HCV awaiting antiviral therapy. European Journal of Clinical Microbiology & Infectious Diseases. 2016;**35**(7):1171-1176. DOI: 10.1007/s10096-016-2650-8

[24] Wang J, Li WT, Zheng YX, Zhao SS, Li N, Huang Y, Zhou RR, Huang ZB, Fan XG. The association between *Helicobacter pylori* infection and chronic hepatitis C: A meta-analysis and trial sequential analysis. Gastroenterology Research and Practice. 2016;**2016**:8780695. DOI: 10.1155/2016/8780695

[25] Sakr SA, Badrah GA, Sheir RA. Histological and histochemical alterations in liver of chronic hepatitis C patients with *Helicobacter pylori* infection. Biomedicine & Pharmacotherapy. 2013;**67**(5):367-374. DOI: 10.1016/j.biopha.2013.03.004

[26] Salehi H, Minakari M, Yaghoutkar A, Tabesh E, Salehi M, Mirbagher L. The effect of *Helicobacter pylori* eradication on liver enzymes in patients referring with unexplained hypertransaminasemia. Advanced Biomedical Research. 2014;**3**:131. DOI: 10.4103/2277-9175.133256

[27] Shapira Y, Agmon-Levin N, Renaudineau Y, Porat-Katz BS, Barzilai O, Ram M, Youinou P, Shoenfeld Y. Serum markers of infections in patients with primary biliary cirrhosis: Evidence of infection burden. Experimental and Molecular Pathology. 2012;**93**(3): 386-390. DOI: 10.1016/j.yexmp.2012.09.012

[28] Casswall TH, Németh A, Nilsson I, Wadström T, Nilsson HO. Helicobacter species DNA in liver and gastric tissues in children and adolescents with chronic liver disease. Scandinavian Journal of Gastroenterology. 2010;**45**(2):160-167. DOI: 10.3109/0036552090 3426915

[29] Dzierzanowska-Fangrat K, Nilsson I, Wozniak M, Jozwiak P, Rozynek E, Woynarowski M, Socha J, Ljungh A, Wadström T. Lack of an association between *Helicobacter* infection and autoimmune hepatitis in children. Polish Journal of Microbiology. 2006;**55**(2):157-159

[30] Toyoda M, Yokomori H, Kaneko F, Yoshida H, Hoshi K, Takeuchi H, Tahara K, Takahashi A, Kudo T, Motoori T, Ohbu M, Kondo H, Hibi T. Primary biliary cirrhosis-autoimmune hepatitis overlap syndrome concomitant with systemic sclerosis, immune thrombocytopenic purpura. Internal Medicine. 2009;**48**(23):2019-2023

[31] Buzás GM. Metabolic consequences of *Helicobacter pylori* infection and eradication. World Journal of Gastroenterology. 2014;**20**(18):5226-5234. DOI: 10.3748/wjg.v20.i18.5226

[32] Castaño-Rodríguez N, Mitchell HM, Kaakoush NO. NAFLD, *Helicobacter* species and the intestinal microbiome. Best Practice & Research. Clinical Gastroenterology. 2017;**31**(6):657-668. DOI: 10.1016/j.bpg.2017.09.008

[33] Sumida Y, Kanemasa K, Imai S, Mori K, Tanaka S, Shimokobe H, Kitamura Y, Fukumoto K, Kakutani A, Ohno T, Taketani H, Seko Y, Ishiba H, Hara T, Okajima A, Yamaguchi K, Moriguchi M, Mitsuyoshi H, Yasui K, Minami M, Itoh Y. *Helicobacter pylori* infection might have a potential role in hepatocyte ballooning in nonalcoholic fatty liver disease. Journal of Gastroenterology. 2015;**50**(9):996-1004. DOI: 10.1007/s00535-015-1039-2

[34] Zhang FM, Yu CH, Chen HT, Shen Z, Hu FL, Yuan XP, Xu GQ. *Helicobacter pylori* infection is associated with gallstones: Epidemiological survey in China. World Journal of Gastroenterology. 2015;**21**(29):8912-8919. DOI: 10.3748/wjg.v21.i29.8912

[35] Sathar SA, Kunnathuparambil SG, Sreesh S, Narayanan P, Vinayakumar KR. *Helicobacter pylori* infection in patients with liver cirrhosis: Prevalence and association with portal hypertensive gastropathy. Annals of Gastroenterology. 2014;**27**(1):48-52

[36] Licinio R, Losurdo G, Carparelli S, Iannone A, Giorgio F, Barone M, Principi M, Ierardi E, Di Leo A. *Helicobacter pylori*, liver cirrhosis, and portal hypertension: An updated appraisal. Immunopharmacol Immunotoxicol. 2016;**38**(6):408-413. doi:10.1080/08923973. 2016.1247855

[37] Pogorzelska J, Łapińska M, Kalinowska A, Łapiński TW, Flisiak R. *Helicobacter pylori* infection among patients with liver cirrhosis. European Journal of Gastroenterology & Hepatology. 2017;**29**(10):1161-1165. DOI: 10.1097/MEG.0000000000000928

[38] Feng H, Zhou X, Zhang G. Association between cirrhosis and *Helicobacter pylori* infection: A meta-analysis. European Journal of Gastroenterology & Hepatology. 2014;**26**:1309-1319. DOI: 10.1097/MEG.0000000000000220

[39] El-Masry S, El-Shahat M, Badra G, Aboel-Nour MF, Lotfy M. *Helicobacter pylori* and hepatitis C Virus coinfection in Egyptian patients. Journal of Global Infectious Diseases. 2010;**2**(1):4-9. DOI: 10.4103/0974-777X.59244

[40] Takashima T, Enomoto H, Iwata Y, Nishikawa H, Yoh K, Hasegawa K, Nakano C, Yuri Y, Ishii N, Miyamoto Y, Takata R, Nishimura T, Ishii A, Sakai Y, Aizawa N, Ikeda N, Iijima H, Nishiguchi S. Effects of *Helicobacter pylori* eradication on the platelet count in hepatitis C virus-infected patients. Journal of Clinical Medical Research. 2016;**8**(12):854-858. DOI: 10.14740/jocmr2725w

[41] Rekha C, Phanidhar S, Sagar AV, Revathi A, Asra WA. Role of *Helicobacter pylori* and hyperammonemia in subclinical hepatic encephalopathy in cirrhosis of liver. Indian Journal of Clinical Biochemistry. 2007;**22**(2):136-139. DOI: 10.1007/BF02913332

[42] Kountouras J, Deretzi G, Zavos C, Katsinelos P. *Helicobacter pylori* infection and liver cirrhosis: Possible association with hepatic encephalopathy and/or post-hepatic encephalopathy cognitive impairment in patients with portal hypertension. Annals of Gastroenterology. 2014;**27**(3):285

[43] Beisele M, Shen Z, Parry N, Mobley M, Taylor NS, Buckley E, Abedin MZ, Dewhirst FE, Fox JG. *Helicobacter marmotae* and novel *Helicobacter* and *Campylobacter* species isolated from the livers and intestines of prairie dogs. Journal of Medical Microbiology. 2011;**60**:1366-1374. DOI: 10.1099/jmm.0.032144-0

[44] Abu Al-Soud W, Stenram U, Ljungh A, Tranberg KG, Nilsson HO, Wadström T. DNA of *Helicobacter* spp. and common gut bacteria in primary liver carcinoma. Digestive and Liver Disease. 2008;**40**(2):126-131. DOI: 10.1016/j.dld.2007.09.011

[45] Ki MR, Goo MJ, Park JK, Hong IH, Ji AR, Han SY, You SY, Lee EM, Kim AY, Park SJ, Lee HJ, Kim SY, Jeong KS. *Helicobacter pylori* accelerates hepatic fibrosis by sensitizing transforming growth factor-β1-induced inflammatory signaling. Laboratory Investigation. 2010;**90**(10):1507-1516. DOI: 10.1038/labinvest.2010.109

[46] Kawanishi S, Ohnishi S, Ma N, Hiraku Y, Oikawa S, Murata M. Nitrative and oxidative DNA damage in infection-related carcinogenesis in relation to cancer stem cells. Genes and Environment. 2017;**38**:26. DOI: 10.1186/s41021-016-0055-7

[47] Boonyanugomol W, Chomvarin C, Sripa B, Bhudhisawasdi V, Khuntikeo N, Hahnvajanawong C, Chamsuwan A. *Helicobacter pylori* in Thai patients with cholangiocarcinoma and its association with biliary inflammation and proliferation. HPB: The Official Journal of the International Hepato Pancreato Biliary Association. 2012;**14**(3):177-184

Section 2

Virulence Factros of *Helicobacter Pylori*

Chapter 5

VacA Genotype in *Helicobacter pylori*

Elnaz Saeidi, Amirhossein Sheikhshahrokh,
Abbas Doosti and Reza Ranjbar

Additional information is available at the end of the chapter

http://dx.doi.org/10.5772/intechopen.81203

Abstract

Helicobacter pylori infection has been recognized as a worldwide problem. *H. pylori* infection is the most prevalent cause of chronic gastritis and has been related to peptic ulcer disease and gastric cancer. It is considered that *H. pylori* infects half of the world's population. Several virulence factors are produced by *H. pylori* in which each of them is related to an increase in the risk of disease development. The vacuolating cytotoxin (VacA) is one of these virulence factors. The first defined action of VacA was induction of intracellular vacuolation. VacA uses a variation in other effects on target cells, such as disruption of mitochondrial functions, stimulation of apoptosis, and blockade of T-cell proliferation, for the induction of vacuolation. In addition, VacA has an important role for colonization of *H. pylori* in vivo.

Keywords: *Helicobacter pylori*, disease, vacuolating cytotoxin (VacA), vacuolation

1. Introduction

Helicobacter pylori (*H. pylori*) is a gram-negative and microaerophilic bacterium, which usually colonizes in the human stomach. *H. pylori* affects about half of the human population worldwide, which exists in their upper gastrointestinal tract. Though all the factors have not been known, we could say that the infection is most likely to happen at a young age and happens more common in developing countries [1]. Prevalence of infection is through human contact mainly via the gastric-oral way [2]. *H. pylori* is related to some diseases such as peptic ulcer disease, gastric ulcers, mucosa-associated lymphoid tissue (MALT) lymphoma, and gastric cancer [3].

Study on this microaerophilic spiral-shaped bacterium is interesting. *H. pylori* is a part of a quickly growing genus. New species are being derived from numerous vertebrate hosts.

In addition, other Helicobacter species are being derived from nongastric parts in humans and might have a role in diseases that formerly had no certain etiologic factor.

H. pylori has polar-sheathed flagella, which helps in motility. In addition, these structures also have a terminal bulb that could make it more adapted to swimming through mucus. Moreover, on the surface, there are special biological characteristics in the lipopolysaccharide, and in order to escape from the host responses, genes that control addition of the O-side chains can phase vary. Moreover, *H. pylori* has a special peptidoglycan structure, which is different from other gram-negative bacteria. Also *H. pylori* releases an autotransported vacuolating cytotoxin that makes the abnormal phenotype of vacuolation in host cells.

For the first time in 1982, two Australian scientists Barry Marshall and Robin Warren identified *H. pylori* in a patient with chronic gastritis and gastric ulcers. Before that, it was not believed to have a microbial reason. By the successful culture of *H. pylori*, a large number of researchers investigated the epidemiology of transmission of the organism. In addition, it is connected with the development of duodenal ulcers and stomach cancer. More than 80% of infected population with the bacterium are asymptomatic, and it might have an important impress in the natural stomach ecology [4]. However, almost 10–15% of people infected with *H. pylori* shown severe gastric disorders containing peptic ulcers, gastric lymphoma, mucosa-associated lymphoid tissue (MALT) lymphoma, and gastric adenocarcinoma [5].

In 1983, *H. pylori* was cultured from human gastric tissue for the first time [5]. After many years, a proteinaceous component known as "vacuolating cytotoxin" was found in *H. pylori* broth culture supernatants. By adding vacuolating cytotoxin to cultured eukaryotic cells, the cells became vacuolated [6]. Formerly, bacterial toxins had not been reported by this function. In the further studies, the identity of the vacuolating toxin was shown [7, 8] and revealed that vacuolating cytotoxin (VacA) has different characteristics and activities compared to other bacterial toxins.

Multiple virulence factors are produced by *H. pylori* in which each of them is related to an increase in the risk of disease extension. Cytotoxin-associated gene A (CagA) and the vacuolating cytotoxin (VacA) are the virulence factors [9].

Infection by *H. pylori* strains including the toxigenic allelic s1 form of VacA increased the risk of peptic ulceration and gastric cancer [10]. VacA was termed because of its ability to cause "vacuole"-like membrane vesicles in the cytoplasm of gastric cells [11]. However, its function in *H. pylori* pathogenesis has not been clear yet. VacA is a pore-forming toxin (PFT). VacA uses a variety of other effects on target cells, such as disruption of mitochondrial functions, stimulation of apoptosis, and blockade of T-cell proliferation, for the induction of vacuolation [12]. In addition, VacA has an important role for colonization of *H. pylori* in vivo [13].

A type IV secretion system is encoded in the cag pathogenicity island (cagPAI), and it replaces CagA into gastric epithelial cells. It causes morphological changes and proinflammatory cytokine secretion [14].

2. *H. pylori*

The gastric mucosa of almost 50% of the world's population has colonized by *Helicobacter pylori* and is related *to gastroduodenal diseases ranging from superficial and chronic gastritis, and duodenal and gastric ulcers to gastric carcinoma and mucosa-associated lymphoid tissue (MALT) lymphoma. [15–17]. There is also some evidence that infection by *H. pylori* may be related in increasing the intensity or risk of infection by other gastrointestinal pathogens and in childhood malnutrition, especially in countries that are less developed [18, 19].

H. pylori was the first bacterial species that is genome sequenced and compared with two independent isolates [20, 21]. Further comparison has presented the first detailed view point at the physical chromosomal organization and has started to recognize a minimal set of common genes that can be considered as candidates for therapeutic strategies.

The two independent *H. pylori* genomes, which are completely sequenced, have different origins. *H. pylori* 26695 was isolated from a patient with gastritis in the United Kingdom in the early 1980s and sequenced by the Institute for Genomic Research [22]. Before sequencing, this strain has been passaged frequently in the laboratory. Also, *H. pylori* J99 was isolated from a patient with a duodenal ulcer and duodenitis in the United States in 1994 and sequenced in a collaborative effort between Astra AB (now AstraZeneca PLC) and Genome Therapeutics Corporation. This strain had not been extensively passaged before sequencing [20]. By using a random shot-gun approach from libraries of cloned chromosomal fragments of ~2.5 kb, J99 and 26695 were sequenced. Like the most microbial genome sequencing projects until now about 45,000 sequence reads in the case of J99 using PHRAP, which resulted in 68 nonredundant contigs, representing almost 98% of the genome, assembled it. PFGE analysis and probe hybridization confirmed the assembly of the *H. pylori* J99 genome [20].

The 26695 genome was 24 kb larger than J99. However, both the J99 and 26695 genomes possessed a total (G + C)% of 39%. There is some similarity in J99 and 26695, such as average lengths of coding sequences, coding density, and the bias of initiation codons. The genome of J99, consistent with the genome of strain 26695, had no clearly recognizable origin of replication. Near the origin of replication in prokaryotes, specific genes, including dnaA, dnaN, and gyrA, are often detected. However, these genes are not in close nearness either to each other or to the repeated heptamer that was determined as nucleotide number one in both published *H. pylori* sequences.

Additional evidence is that this position may be regarded as the replication origin achieved from using an algorithm that analyzes the bias of short oligomers whose direction is preferentially skewed around the replication origin of prokaryotes [23].

Leunk et al. found massive vacuolar degeneration of various cultured epithelial cell lines in supernatants from broth cultures of *Helicobacter pylori*, in 1988 [6]. After that, numerous studies throughout the world have done on the nature of this toxic activity and its effect in *H. pylori*-induced disease. The protein mediating the effect was purified and called the vacuolating cytotoxin in 1992 [7]. In 1994, discovery of the amino-terminal sequence of the protein led to the cloning and sequencing of the toxin gene, which was nominated vacA [8, 24–26].

Subsequent to the primary characterization of the toxin and its gene, research has focused on VacA structure, the mechanisms underlying VacA's toxic activity, naturally happening differences between VacA proteins produced by various strains of *H. pylori*, and the clinical significance of VacA polymorphism.

Studies on VacA has expanded, not only because of its potential as a novel tool for exploring features of eukaryotic cell biology but also mainly because of its supposed function in the pathogenesis of *H. pylori*-related diseases, in specific peptic ulceration and distal gastric adenocarcinoma. The accurate function of VacA in these diseases is still under research, but VacA may contribute to the capacity of *H. pylori* to colonize and persist in the human gastric mucosa and may also contribute immediately to gastric epithelial damage. Therefore, VacA is a purpose for therapeutic intervention and a candidate for inclusion in a vaccine against *H. pylori*.

3. The *vacA* gene

Vacuolating cytotoxin (vacA) is the most commonly identified virulence factor among *H. pylori* strains. VacA belongs to the group of genes with mutable genotypes related to damage to gastric epithelial cells. This gene exists in almost all strains of *H. pylori*. This gene is polymorphic and contains variable signal regions (type s1 or type s2) and midregions (type m1 or type m2) [27] and intermediate regions (i1 and i2 alleles, and the rare i3 allele) [28] (**Figure 1**). There are various levels of its cytotoxicity that is caused by the variety of signal (s) and mid (m) regions of vacA gene [29]. S region variations are more related to the vacuolating activity of vacA, and m region variations have effect on binding of the toxin to the host cells, as reasons are contributed to define cell specificity [29].

A copy of the toxin gene, vacA, exists in all *H. pylori* strains. The vacA transcript is monocistronic. Transcriptional start point in this gene is located about 119 nucleotides upstream from the ATG start codon [25, 30]. The capacity of *H. pylori* to induce vacuolation in epithelial cells abrogates by insertional mutagenesis of vacA. In addition, it interrupts a number of other vacA-induced toxic effects [24, 25, 31]. Alleles of vacA from about 25 different *H. pylori* strains have been sequenced and range from 3864 to 3933 nucleotides in length [8, 24, 32–34].

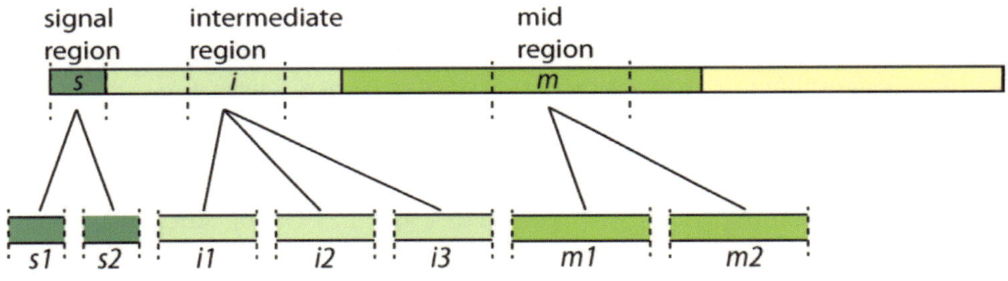

Figure 1. VacA gene.

4. The association of vacA types with cytotoxin production

Significant genetic diversity exists between vacA alleles from different strains. These alleles could be classified into various families. The most significant studied form of VacA is encoded by type s1/m1 vacA alleles. s1/m1 vacA alleles typically encode VacA proteins, related to a high level of vacuolating cytotoxin activity [32], while s1/m2 strains have moderate toxin production [29]. In fact, s1/m2 strains that have an i1 allele are able to induce vacuolation. However, s1/m2 strains that have an i2 allele are not able to induce vacuolation [28]. s2/m2 strains have rare or even absent toxin production [29].

By comparing vacA s1 and m1 strains with vacA s2 and m2 strains, which are less virulent, it was revealed that *H. pylori* vacA s1 and m1 strains are related to higher levels of inflammation in the gastric mucosa and increased risk for gastric atrophy and carcinoma. After the explanation of the vacA i-region, it was also revealed that the determinant of cytotoxicity i1 allele is related to gastric carcinoma [28].

5. VacA proteins

VacA encodes a protein with a mass of about 140 kDa; however, under denaturing conditions, the mature secreted VacA toxin drifts as a band of almost 90 kDa [7, 8, 24–26]. A comparison of the amino-terminal sequence of the mature secreted toxin with that predicted for the protoxin shows that a 33-amino-acid amino-terminal signal sequence is cleaved during the procedure of VacA secretion. Investigations by antisera raised against various regions of recombinant VacA show that a polypeptide of about 33 kDa isolated from the carboxy-terminal portion of the protoxin stays localized to the bacteria and is not secreted [26]. This carboxy-terminal portion of VacA seems to contain amphipathic β-sheets capable of forming a β-barrel structure and has a terminal phenylalanine-containing motif that is available in several outer membrane proteins [25]. These qualities, with a pair of cysteine residues nearly the carboxy-terminus of the mature secreted protein, are specification of a family of secreted bacterial proteins called autotransporters [35]. Autotransporters for export across the bacterial outer membrane do not need any auxiliary proteins. By studying the Neisseria gonorrhoeae IgA1 protease, we achieved this information of autotransporter export. Translocation of IgA1 protease through the bacterial cytoplasmic membrane is achieved via a Sec-mediated process and is accompanied by cleavage of an amino-terminal signal peptide. After inserting the carboxy-terminal β-barrel domain into the outer membrane, it functions as a pore through that the residue of the molecule passes. The mature secreted IgA1 protease is produced by autoproteolytic cleavage. The carboxy-terminal domain stays related to the outer membrane [36].

Primary studies showed, despite mature VacA monomers are about 90 kDa in mass, that the toxin exists as a much larger complex or aggregate under nondenaturing conditions [7]. Lupetti et al. investigated the ultrastructure of purified VacA using deep-etch electron microscopy. They illustrated that the toxin forms into large flower-shaped complexes that appear to consist of a central ring surrounded by six or seven "petals" [37]. An accurate view of the surface of VacA oligomers is presented by three-dimensional reconstructions of these deep-etch metal copies (**Figure 2**) [38]. Moreover, the classical flower-like complexes, VacA, could be assembled into other type of complex, that is named a "flat form," which includes of six or seven petals without a notable central ring [37, 39].

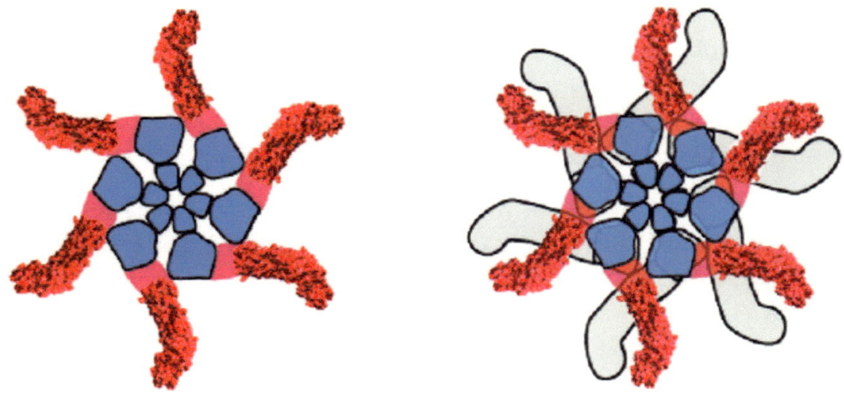

Figure 2. VacA oligomer.

The petals that contain the flat form generally radiate from the center of the complex with a specific clockwise chirality. Several models have suggested clarifying the assembly of VacA into flower-like complexes and chiral flat forms. In one of these models, the flower-like forms are proposed to contain six or seven monomers of about 90 kDa [38, 39]. In another model, the flower-like forms are considered to be dodecamers or tetradecamers of VacA monomers of about 90 kDa, and flat forms are proposed to be hexamers or heptamers [37]. In acidic or alkaline pH, VacA oligomers separate into monomeric parts of approximately 90 kDa, each measuring of about 6 by 14 nm [37, 40, 41]. This pH-mediated disassembly is related by a marked enhance in VacA cytotoxic activity [39, 41–43].

This opinion exists that VacA monomers have more cytotoxic activity than water-soluble VacA oligomers. Subsequent researches about VacA structure have been undertaken using atomic force microscopic imaging of purified toxin bound to supported lipid bilayers [44].

Two-dimensional crystalline arrays of VacA on lipid bilayers include an arranged array of hexagonal central rings connected by thin connectors to peripheral domains.

In-frame deletions in the portion of VacA encoding the amino-terminal region of the toxin produced mutant strains of *H. pylori*. Mutant strains of *H. pylori* express truncated VacA proteins. These proteins are secreted, though fail to oligomerize and lack recognizable cytotoxic activity [45, 46]. VacA Δ91–330 is a mutant VacA protein that has water-soluble dimeric form, which has an ultrastructural appearance similar to that of the peripheral petals of VacA oligomers [45]. The peripheral petals of VacA oligomers can be consistent with the carboxy-terminal portion of the mature secreted VacA polypeptide.

6. Functional domains in VacA

The purified ~90 kDa VacA toxin through extended storage or incubation with trypsin break down into ~37 and ~58 kDa components, which are isolated from the amino terminus and carboxy terminus of the protein. Proteolytic cleavage occurs at a site containing multiple charged amino acids [26]. In fact, the 37 and 58 kDa fragments of VacA are considered as subunits or domains of the holotoxin [47].

Burroni et al. manufactured an *H. pylori* mutant in order to specify whether cleavage of VacA into 37 and 58 kDa fragments is needed for toxin activity. In this *H. pylori* mutant, the region of vacA encoding the 46 amino acids flanking the VacA cleavage site was removed [48]. Because of this fact that this mutant VacA was entirely active, it is informing that cleavage of the exposed loop is not required for activity. While the wild-type VacA produced by the parent strain prefer to form seven-sided complexes, in contrast, the mutant prefers to form six-sided complexes. This revealed that deleting the exposed loop presented structural restriction.

In experiments wherever mutant forms of vacA under the control of a eukaryotic promoter have been expressed from plasmids in the cytosol of epithelial cells, it has been explained that the minimal region of VacA is required for vacuolating activity [11]. These experiments revealed that the epithelial cell lines, which transfected with plasmid constructs encoding either the full-length ~90 kDa secreted toxin or amino- or carboxy-terminally truncated fragments. In addition, these experiments represented that a VacA protein lacking most of the carboxy-terminal 58 kDa domain preserved complete vacuolating activity [49, 50]. By eliminating 10 amino acids from the amino-terminus, activity was entirely abolished, and by eliminating 6 amino acids from the amino-terminus, activity was only in part abolished [49, 50]. The minimal VacA domain that presented complete vacuolating activity when expressed intracellularly was a peptide containing amino acids 1–422, which is the 37-kDa domain plus a fragment of the 58-kDa domain [50]. The 37 kDa fragment was inactive in alone, but coexpression of this fragment with a fragment including the amino-terminal 165 amino acids of the 58 kDa fragment resulted in complete vacuolating activity [50]. A conceivable explanation for the importance of the VacA amino terminus was determined by hydrophobicity plots. In fact, the only hydrophobic region in VacA is amino acids 1–32 of this region, and it is long enough to span a membrane.

A *H. pylori* vacA partial deletion mutant was produced, which lacked codons for amino acids 6–27, for more research [51]. The structure of mutant VacA did not have variations compared to wild-type VacA; however, mutant VacA lacked cytotoxic activity.

In addition, alanine scanning mutagenesis showed that point mutations at proline 9 or glycine 14 entirely abrogated VacA activity [52]. Another factor that abrogated toxin activity is the addition of an amino-terminal hydrophilic extension to VacA [53]. As a result, it is obvious that the amino-terminal hydrophobic region has an important role in toxin activity.

7. Receptor binding region

It is demonstrated that amino acid sequences located in the carboxy-terminal portion of the mature protein mediated binding of VacA to cells.

Investigations on the purified 58 kDa fragment from a mutant *H. pylori* strain represent that this protein binds to HeLa cells with kinetics similar to those of the intact toxin [45]. The binding of VacA to cells is inhibited by polyclonal antiserum reactive with the 58 kDa domain [54]. Several natural forms of VacA have significant divergent amino acid sequences in the 58 kDa domain, which are called m2 forms. These forms cause vacuolation in a more confined range of cultured epithelial cell lines. Differences in cell binding would be a reason for this [34]. VacA with a type m2 58 kDa domain, that did not cause HeLa cell vacuolation when applied externally, affected vacuolation when expressed from a plasmid in the HeLa cell cytoplasm. This indicates that m2 VacA is entirely active but cannot get to its site of action. This would be because of inability to bind to the cell [11].

Investigations by naturally occurring and engineered m1/m2 chimeric proteins [55] propose that an ~40 amino acid region near the amino-terminal end of the 58 kDa domain is required for HeLa cell vacuolation and can have a role in HeLa cell binding.

8. Activity of VacA

Epithelial cell vacuolation in vitro occurs by VacA; however, this does not cause cell death quickly. Cell death in human gastric epithelial cells that are exposed to high doses of toxin is reported after 2 days [56]. On the other hand, cell death does not normally happen in immortalized cell lines exposed to the toxin. As an example, incubation of AZ-521 gastric epithelial cells with VacA for several hours causes decreased mitochondrial ATP production and decreased oxygen utilization but does not result in cell death [57].

The exact mechanisms of binding and uptake of VacA by cells are not clearly understood yet. The prototypic s1/m1 form of VacA binds to HeLa cells in a saturable manner recognized by flow-cytometry analysis [58]. However, saturable binding has not been indicated with classical ligand binding assays with 12I-labeled VacA [43]. Activation of VacA by acid treatment significantly increases its vacuolating activity but does not remarkably increase its binding to HeLa or Baby Hamster Kidney (BHK) cells [40, 58]. However, binding of the toxin to the gastric cell line AZ-521 is increased by acid activation [41]. A number of specific VacA receptors were proposed. Activated VacA binds to a 250 kDa receptor protein-tyrosine phosphatase β (RPTPβ) in the AZ-521 system that regulates intracellular tyrosine phosphorylation [41, 59].

Autotransporters are a family of secreted bacterial proteins, which are determined by mentioned features, together with a pair of cysteine residues near the carboxy-terminus of the mature secreted protein. It is suggested that RPTPβ has an important role in binding VacA to cells and following intoxication. Treatment of the HL-60 cell line with phorbol 12-myristate 13-acetate (PMA) causes stimulation of RPTPβ expression that is occurred with stimulation of VacA sensitivity [60]. BHK-21 cells are insensitive to VacA, but transfection with expression vectors including the RPTPβ gene can make them sensitive. Antisense oligonucleotides in PMA-treated HL-60 cells lead to ablation of RPTPβ synthesis. As a result, a considerable reduction occurs in VacA-induced vacuolation. An unidentified 140 kDa protein in AZ-521 and AGS cells and the epidermal growth factor receptor in HeLa cells [61, 62] are two other particular VacA receptors. These evidence suggested that multiple surface-binding sites recognized by both inactive and activated VacA exist; in addition, specific VacA receptors exist that are variably expressed in different cell lines.

Both 58 and 37 kDa regions are needed for VacA internalization [45]. VacA should be preactivated by disposal of acid or alkali, in order to be internalized [43]. Internalization happens through an energy-dependent process; the exact nature of which is not clear. However, it may be a receptor-mediated endocytosis. VacA molecules localize in membrane vesicles, after internalization [54]. Then localized VacA molecules are transported along the endocytic pathway to vacuolar-type (V-) ATPase-positive late endosomes and lysosomes. In this state, they accumulate and persist for some days [63, 64].

The first defined action of VacA was induction of intracellular vacuoles [64, 65]. The vacuolar membranes include both late endosomal and lysosomal markers, indicating that the vacuoles are derived from these sections [66, 67].

The complete activity of V-ATPase and the existence of weak bases are needed for the formation of VacA-induced vacuoles, which indicated that vacuoles are derived from the accumulation of weak bases within acidic sections, and with water influx and swelling followed [63, 64, 68, 69]. Moreover, the membrane traffic regulator rab7 and the actin-cytoskeleton-associated Rac1 are two small GTP-binding proteins that involved in vacuole biogenesis [70, 71]. Rac1 and rab7 are related with the membrane of VacA-induced vacuoles. The expression of rab7 or Rac1 dominant negative mutants inhibits vacuolization, and the expression of rab7 or Rac1 dominant positive mutants potentiates vacuolization. It has been proposed that membrane fusion events and the cytoskeleton supporting late endosomal sections regulated vacuole development. VacA destructs the transport of acidic hydrolases to lysosomes and causes the release of these enzymes into the extracellular medium in HeLa cells [72]. VacA caused decrement of the degradative power of HeLa cell lysosomes and also decrement of the antigen-processing compartment of B lymphocytes [72, 73].

VacA is unable to vacuolate epithelial monolayers of MDCK I, T84, or epH4 cells on porous filters. In addition, MDCK I, T84, or epH4 cells do not show signs of endolysosomal dysfunction [74].

Subsequently, disposal to VacA, transepithelial electrical resistance (TER) reduces, occurred with an increase in transepithelial flux of low-molecular-weight molecules [74]. There are some reasons, which propose that VacA modulates the resistance of these model epithelia through a paracellular effect. These reasons include the size selectivity of this increased epithelial permeation, lack of accompanying vacuolation, and lack of redistribution of junctional proteins. Just epithelial cell monolayers capable of expanding a TER higher than 1000–1200 Ω/cm^2 are affected. By utilizing the isogenic mutant strains, this is confirmed that the effect is dependent on VacA [31]. In MDCK cells, m2 type of VacA decreases TER. However, it does not lead to vacuolation in this cell line even when cells are nonconfluent [31]. It is corroborated that vacuolation and increased permeability of monolayers are separate and independent effects.

VacA constructs ion channels in model lipid bilayers and cell plasma membranes. This occurrence may underlie all the other consequences of VacA. Acidic conditions cause disassembly of the inactive VacA oligomer, which permits insertion of the toxin into lipid bilayers [66, 73, 75].

Investigations with planar model membranes represent that membrane insertion is followed by the formation of voltage-dependent, low-conductance (10–30 pS in 2 M KCl), and anion-selective channels [76, 77].

Patch clamp analysis of HeLa cells indicates that VacA forms plasma membrane channels with features similar to those perceived in model membranes [78]. Different anion channel blockers inhibit VacA channels in vitro with various powers and are able to prevent and partially inverse vacuolation of HeLa cells [78, 79], informing an essential role of the anion channel in vacuolation [41]. With permitting anions to permeate into late endosomes, the endocytosed VacA channel increases the turnover of the electrogenic V-ATPase that causes accumulation of weak bases and leads to vacuole formation by water influx [80, 81]. Because of that, internalization of surface-bound VacA is required for the further development of vacuolation; this hypothesis is acceptable [43]. Vacuolation in this model can be considered as a side effect of the massive accumulation of endocytosed VacA channels in endolysosomes. With 5-nitro-2-(3-phenylpropylamine) benzoic acid (NPPB), VacA epithelial permeabilization of MDCK I cells can be partly prevented and reversed, the most efficient blocker of VacA channels, implying that epithelial permeabilization, similar to vacuolation, is less important for the formation of apical anion channels [78]. VacA induces an increased apical anion secretion in Caco-2 cells, and this also is blocked by NPPB [82], implying that it is also because of VacA anion channel formation.

9. Discussion

H. pylori is considered as a significant cause of chronic active gastritis, peptic ulcer, and atrophic gastritis. It is related to an enhanced risk of gastric adenocarcinoma and mucosa-associated lymphoid tissue (MALT). VacA is a virulence factor related to peptic ulcer. Moreover, oral administration of VacA leads to gastric mucosal damage in mice. It could be concluded that VacA might contribute to epithelial cell damage or peptic ulceration in H. pylori-infected humans [83].

A toxin that has damaging outcomes on epithelial cells is produced by *H. pylori*. In addition, colonization of *H. pylori* has contributed in the development of peptic ulceration. By this information, it could be concluded that VacA directly harms the gastric and duodenal epithelium in vivo; therefore, it leads to ulcers. *H. pylori* strains which have vacuolating cytotoxin activity in vitro are more often associated with disease than *H. pylori* strains which are noncytotoxic strains. There is a significant association between vacuolating activity and peptic ulcer disease, which is revealed by some studies all over the world [32, 84–86]; however, this association is not true in all situations. Often noncytotoxic *H. pylori* isolates from patients with peptic ulceration, and cytotoxic *H. pylori* isolates from patients without peptic ulceration. However, explanation of these studies depends on some factors.

There have been many studies on the relationship between specific vacA genotypes and diseases, which are developed by multiple vacA genotypes and explained by polymerase chain reaction (PCR)-based methodology for discrimination between them [32, 81, 87]. It was proved by most of these studies from outside Asia that s1 strains are more often associated with peptic ulceration or gastric carcinoma than s2 strains [32, 81, 88–91].

H. pylori vacA s1 and m1 strains are related to higher levels of inflammation in the gastric mucosa and increased risk for gastric atrophy and carcinoma. In addition, it was revealed that the determinant of cytotoxicity i1 allele is related to gastric carcinoma. So, evaluation of characterization of this region as a determinant of the clinical outcome of *H. pylori* infection could be used [28].

10. Conclusion

Helicobacter pylori has been investigated since its first culture in 1982 from a gastric biopsy. Cytotoxin-associated gene A (CagA) and the vacuolating cytotoxin (VacA) are the virulence factors which are produced by *H. pylori* and are related to an increase in the risk of disease extension [9].

VacA has a significant role in the pathogenesis of *H. pylori*-associated diseases, especially in peptic ulceration and distal gastric adenocarcinoma. For this reason, VacA has been studied widely. VacA is still under examination in order to find out its accurate role in these diseases. However, VacA makes *H. pylori* able to colonize in the human gastric mucosa. Moreover, VacA could have a role in gastric epithelial damage. For this reason, VacA is a target for therapeutic intervention and also is considered for usage in a vaccine against *H. pylori*.

Author details

Elnaz Saeidi[1], Amirhossein Sheikhshahrokh[1*], Abbas Doosti[2] and Reza Ranjbar[3]

*Address all correspondence to: amirhossein.sheikhshahrokh@yahoo.com

1 Young Researcher Club, Islamic Azad University, Shahrekord, Iran

2 Biotechnology Research Center, Shahrekord Branch, Islamic Azad University, Shahrekord, Iran

3 Molecular Biology Research Center, Systems Biology and Poisonings Institute, Baqiyatallah University of Medical Sciences, Tehran, Iran

References

[1] Control, C.F.D. and Prevention. CDC Health Information for International Travel 2014: The Yellow Book. Oxford, United Kingdom: Oxford University Press; 2013

[2] Amieva MR, El-Omar EM. Host-bacterial interactions in *Helicobacter pylori* infection. Gastroenterology. 2008;**134**(1):306-323

[3] Garza-González E et al. A review of *Helicobacter pylori* diagnosis, treatment, and methods to detect eradication. World Journal of Gastroenterology. 2014;**20**(6):1438

[4] Blaser MJ. Who are we?: Indigenous microbes and the ecology of human diseases. EMBO Reports. 2006;**7**(10):956-960

[5] Marshall B, Warren JR. Unidentified curved bacilli in the stomach of patients with gastritis and peptic ulceration. The Lancet. 1984;**323**(8390):1311-1315

[6] Leunk R et al. Cytotoxic activity in broth-culture filtrates of *Campylobacter pylori*. Journal of Medical Microbiology. 1988;**26**(2):93-99

[7] Cover TL, Blaser M. Purification and characterization of the vacuolating toxin from *Helicobacter pylori*. Journal of Biological Chemistry. 1992;**267**(15):10570-10575

[8] Phadnis SH et al. Pathological significance and molecular characterization of the vacuolating toxin gene of *Helicobacter pylori*. Infection and Immunity. 1994;**62**(5):1557-1565

[9] Kusters JG, van Vliet AH, Kuipers EJ. Pathogenesis of *Helicobacter pylori* infection. Clinical Microbiology Reviews. 2006;**19**(3):449-490

[10] Louw J et al. The relationship between *Helicobacter pylori* infection, the virulence genotypes of the infecting strain and gastric cancer in the African setting. Helicobacter. 2001;**6**(4):268-273

[11] De Bernard M et al. *Helicobacter pylori* toxin VacA induces vacuole formation by acting in the cell cytosol. Molecular Microbiology. 1997;**26**(4):665-674

[12] Cover TL, Blanke SR. *Helicobacter pylori* VacA, a paradigm for toxin multifunctionality. Nature Reviews Microbiology. 2005;**3**(4):320

[13] Salama NR et al. Vacuolating cytotoxin of *Helicobacter pylori* plays a role during colonization in a mouse model of infection. Infection and Immunity. 2001;**69**(2):730-736

[14] Noto JM, Peek RM. The *Helicobacter pylori* cag pathogenicity island. In: Helicobacter Species. Springer; 2012. pp. 41-50

[15] Cover TL, Blaser MJ. *Helicobacter pylori* and gastroduodenal disease. Annual Review of Medicine. 1992;**43**(1):135-145

[16] Hunt R. The role of *Helicobacter pylori* in pathogenesis: The spectrum of clinical outcomes. Scandinavian Journal of Gastroenterology. 1996;**31**(sup 220):3-9

[17] Labigne A. Determinants of *Helicobacter pylori* pathogenicity. Infectious Agents and Disease. 1996;**5**(4):191-202

[18] Clemens J et al. Impact of infection by *Helicobacter pylori* on the risk and severity of endemic cholera. Journal of Infectious Diseases. 1995;**171**(6):1653-1656

[19] Dale A et al. *Helicobacter pylori* infection, gastric acid secretion, and infant growth. Journal of Pediatric Gastroenterology and Nutrition. 1998;**26**(4):393-397

[20] Alm RA et al. Genomic-sequence comparison of two unrelated isolates of the human gastric pathogen *Helicobacter pylori*. Nature. 1999;**397**(6715):176

[21] Tomb JF et al. The complete genome sequence of the gastric pathogen *Helicobacter pylori*. Nature. 1997;**388**(6642):539-547

[22] Eaton KA, Morgan D, Krakowka S. *Campylobacter pylori* virulence factors in gnotobiotic piglets. Infection and Immunity. 1989;**57**(4):1119-1125

[23] Salzberg SL et al. Skewed oligomers and origins of replication. Gene. 1998;**217**(1):57-67

[24] Cover TL et al. Divergence of genetic sequences for the vacuolating cytotoxin among *Helicobacter pylori* strains. Journal of Biological Chemistry. 1994;**269**(14):10566-10573

[25] Schmitt W, Haas R. Genetic analysis of the *Helicobacter pylori* vacuoiating cytotoxin: Structural similarities with the IgA protease type of exported protein. Molecular Microbiology. 1994;**12**(2):307-319

[26] Telford JL et al. Gene structure of the *Helicobacter pylori* cytotoxin and evidence of its key role in gastric disease. Journal of Experimental Medicine. 1994;**179**(5):1653-1658

[27] Saeidi E, Sheikhshahrokh A. VacA genotype status of *Helicobacter pylori* isolated from foods with animal origin. BioMed Research International. 2016;**2016**

[28] Ferreira RM et al. A novel method for genotyping *Helicobacter pylori* vacA intermediate region directly in gastric biopsy specimens. Journal of Clinical Microbiology. 2012. DOI: 10.1128/JCM.02087-12

[29] Sheikh AF et al. CagA and vacA allelic combination of *Helicobacter pylori* in gastroduodenal disorders. Microbial Pathogenesis. 2018;**122**:144-150

[30] Forsyth M et al. Heterogeneity in levels of vacuolating cytotoxin gene (vacA) transcription among *Helicobacter pylori* strains. Infection and Immunity. 1998;**66**(7):3088-3094

[31] Pelicic V et al. *Helicobacter pylori* VacA cytotoxin associated with the bacteria increases epithelial permeability independently of its vacuolating activity. Microbiology. 1999;**145**(8): 2043-2050

[32] Atherton JC et al. Mosaicism in vacuolating cytotoxin alleles of *Helicobacter pylori* association of specific vacA types with cytotoxin production and peptic ulceration. Journal of Biological Chemistry. 1995;**270**(30):17771-17777

[33] Ito Y et al. Analysis and typing of the vacA gene from cagA-positive strains of *Helicobacter pylori* isolated in Japan. Journal of Clinical Microbiology. 1997;**35**(7):1710-1714

[34] Pagliaccia C et al. The m2 form of the *Helicobacter pylori* cytotoxin has cell type-specific vacuolating activity. Proceedings of the National Academy of Sciences. 1998;**95**(17): 10212-10217

[35] Henderson IR, Navarro-Garcia F, Nataro JP. The great escape: Structure and function of the autotransporter proteins. Trends in Microbiology. 1998;**6**(9):370-378

[36] Pohlner J et al. Gene structure and extracellular secretion of Neisseria gonorrhoeae IgA protease. Nature. 1987;**325**(6103):458

[37] Lupetti P et al. Oligomeric and subunit structure of the *Helicobacter pylori* vacuolating cytotoxin. The Journal of Cell Biology. 1996;**133**(4):801-807

[38] Lanzavecchia S et al. Three-dimensional reconstruction of metal replicas of the *Helicobacter pylori* vacuolating cytotoxin. Journal of Structural Biology. 1998;**121**(1):9-18

[39] Cover TL, Hanson PI, Heuser JE. Acid-induced dissociation of VacA, the *Helicobacter pylori* vacuolating cytotoxin, reveals its pattern of assembly. The Journal of Cell Biology. 1997;**138**(4):759-769

[40] Molinari M et al. The acid activation of *Helicobacter pylori* toxin VacA: Structural and membrane binding studies. Biochemical and Biophysical Research Communications. 1998;**248**(2):334-340

[41] Yahiro K et al. Activation of *Helicobacter pylori* VacA toxin by alkaline or acid conditions increases its binding to a 250-kDa receptor protein-tyrosine phosphatase β. Journal of Biological Chemistry. 1999;**274**(51):36693-36699

[42] de Bernard M et al. Low pH activates the vacuolating toxin of *Helicobacter pylori,* which becomes acid and pepsin resistant. Journal of Biological Chemistry. 1995;**270**(41): 23937-23940

[43] McClain MS et al. Acid activation of *Helicobacter pylori* vacuolating cytotoxin (VacA) results in toxin internalization by eukaryotic cells. Molecular Microbiology. 2000;**37**(2): 433-442

[44] Czajkowsky DM et al. The vacuolating toxin from *Helicobacter pylori* forms hexameric pores in lipid bilayers at low pH. Proceedings of the National Academy of Sciences. 1999;**96**(5):2001-2006

[45] Reyrat J-M et al. 3D imaging of the 58 kda cell binding subunit of the *Helicobacter pylori* cytotoxin1. Journal of Molecular Biology. 1999;**290**(2):459-470

[46] Vinion-Dubiel AD et al. A dominant negative mutant of *Helicobacter pylori* vacuolating toxin (VacA) inhibits VacA-induced cell vacuolation. Journal of Biological Chemistry. 1999;**274**(53):37736-37742

[47] Nguyen VQ, Caprioli RM, Cover TL. Carboxy-terminal proteolytic processing of *Helicobacter pylori* vacuolating toxin. Infection and Immunity. 2001;**69**(1):543-546

[48] Burroni D et al. Deletion of the major proteolytic site of the *Helicobacter pylori* cytotoxin does not influence toxin activity but favors assembly of the toxin into hexameric structures. Infection and Immunity. 1998;**66**(11):5547-5550

[49] de Bernard M et al. Identification of the *Helicobacter pylori* VacA toxin domain active in the cell cytosol. Infection and Immunity. 1998;**66**(12):6014-6016

[50] Ye D, Willhite DC, Blanke SR. Identification of the minimal intracellular vacuolating domain of the *Helicobacter pylori* vacuolating toxin. Journal of Biological Chemistry. 1999;**274**(14):9277-9282

[51] Weel JF et al. The interrelationship between cytotoxin-associated gene A, vacuolating cytotoxin, and *Helicobacter pylori*-related diseases. Journal of Infectious Diseases. 1996;**173**(5): 1171-1175

[52] Ye D, Blanke SR. Mutational analysis of the *Helicobacter pylori* vacuolating toxin amino terminus: Identification of amino acids essential for cellular vacuolation. Infection and Immunity. 2000;**68**(7):4354-4357

[53] Letley D, Atherton J. Natural diversity in the N terminus of the mature vacuolating cytotoxin of *Helicobacter pylori* determines cytotoxin activity. Journal of Bacteriology. 2000;**182**(11):3278-3280

[54] Garner JA, Cover TL. Binding and internalization of the *Helicobacter pylori* vacuolating cytotoxin by epithelial cells. Infection and Immunity. 1996;**64**(10):4197-4203

[55] Ji X et al. Cell specificity of *Helicobacter pylori* cytotoxin is determined by a short region in the polymorphic midregion. Infection and Immunity. 2000;**68**(6):3754-3757

[56] Smoot D et al. Effects of *Helicobacter pylori* vacuolating cytotoxin on primary cultures of human gastric epithelial cells. Gut. 1996;**39**(6):795-799

[57] Kimura M et al. Vacuolating cytotoxin purified from *Helicobacter pylori* causes mitochondrial damage in human gastric cells. Microbial Pathogenesis. 1999;**26**(1):45-52

[58] Massari P et al. Binding of the *Helicobacter pylori* vacuolating cytotoxin to target cells. Infection and Immunity. 1998;**66**(8):3981-3984

[59] Padilla PI et al. Morphologic differentiation of HL-60 cells is associated with appearance of RPTPβ and induction of *Helicobacter pylori* VacA sensitivity. Journal of Biological Chemistry. 2000;**275**(20):15200-15206

[60] de Bernard M et al. TPA and butyrate increase cell sensitivity to the vacuolating toxin of *Helicobacter pylori*. FEBS Letters. 1998;**436**(2):218-222

[61] Yahiro K et al. *Helicobacter pylori* vacuolating cytotoxin binds to the 140-kDa protein in human gastric cancer cell lines, AZ-521 and AGS. Biochemical and Biophysical Research Communications. 1997;**238**(2):629-632

[62] Seto K et al. Vacuolation induced by cytotoxin from *Helicobacter pylori* is mediated by the EGF receptor in HeLa cells. FEBS Letters. 1998;**431**(3):347-350

[63] Ricci V et al. *Helicobacter pylori* vacuolating toxin accumulates within the endosomal-vacuolar compartment of cultured gastric cells and potentiates the vacuolating activity of ammonia. The Journal of Pathology. 1997;**183**(4):453-459

[64] Sommi P et al. Persistence of *Helicobacter pylori* VacA toxin and vacuolating potential in cultured gastric epithelial cells. American Journal of Physiology - Gastrointestinal and Liver Physiology. 1998;**275**(4):G681-G688

[65] Cover TL et al. Potentiation of *Helicobacter pylori* vacuolating toxin activity by nicotine and other weak bases. Journal of Infectious Diseases. 1992;**166**(5):1073-1078

[66] Moll G et al. Lipid interaction of the 37-kDa and 58-kDa fragments of the *Helicobacter Pylori* cytotoxin. The FEBS Journal. 1995;**234**(3):947-952

[67] Papini E et al. Cellular vacuoles induced by *Helicobacter pylori* originate from late endosomal compartments. Proceedings of the National Academy of Sciences. 1994;**91**(21): 9720-9724

[68] Cover T, Reddy L, Blaser M. Effects of ATPase inhibitors on the response of HeLa cells to *Helicobacter pylori* vacuolating toxin. Infection and Immunity. 1993;**61**(4):1427-1431

[69] Papini E et al. Bafilomycin A1 inhibits *Helicobacter pylori*-induced vacuolization of HeLa cells. Molecular Microbiology. 1993;**7**(2):323-327

[70] Hotchin NA, Cover TL, Akhtar N. Cell vacuolation induced by the VacA cytotoxin of *Helicobacter pylori* is regulated by the Rac1 GTPase. Journal of Biological Chemistry. 2000;**275**(19):14009-14012

[71] Marshall DG et al. Lack of a relationship between Lewis antigen expression and cagA, CagA, vacA and VacA status of Irish *Helicobacter pylori* isolates. FEMS Immunology and Medical Microbiology. 1999;**24**(1):79-90

[72] Satin B et al. Effect of *Helicobacter pylori* vacuolating toxin on maturation and extracellular release of procathepsin D and on epidermal growth factor degradation. Journal of Biological Chemistry. 1997;**272**(40):25022-25028

[73] Molinari M et al. Selective inhibition of Ii-dependent antigen presentation by *Helicobacter pylori* toxin VacA. Journal of Experimental Medicine. 1998;**187**(1):135-140

[74] Papini E et al. Selective increase of the permeability of polarized epithelial cell monolayers by *Helicobacter pylori* vacuolating toxin. The Journal of Clinical Investigation. 1998;**102**(4):813-820

[75] Pagliaccia C et al. Structure and interaction of VacA of *Helicobacter pylori* with a lipid membrane. The FEBS Journal. 2000;**267**(1):104-109

[76] Iwamoto H et al. VacA from *Helicobacter pylori*: A hexameric chloride channel. FEBS Letters. 1999;**450**(1-2):101-104

[77] Tombola F et al. *Helicobacter pylori* vacuolating toxin forms anion-selective channels in planar lipid bilayers: Possible implications for the mechanism of cellular vacuolation. Biophysical Journal. 1999;**76**(3):1401-1409

[78] Szabò I et al. Formation of anion-selective channels in the cell plasma membrane by the toxin VacA of *Helicobacter pylori* is required for its biological activity. The EMBO Journal. 1999;**18**(20):5517-5527

[79] Tombola F et al. Inhibition of the vacuolating and anion channel activities of the VacA toxin of *Helicobacter pylori*. FEBS Letters. 1999;**460**(2):221-225

[80] Tee W, Lambert JR, Dwyer B. Cytotoxin production by *Helicobacter pylori* from patients with upper gastrointestinal tract diseases. Journal of Clinical Microbiology. 1995;**33**(5):1203-1205

[81] van Doorn LJ et al. Clinical relevance of the cagA, vacA, and iceA status of *Helicobacter pylori*. Gastroenterology. 1998;**115**(1):58-66

[82] Guarino A et al. Enterotoxic effect of the vacuolating toxin produced by *Helicobacter pylori* in Caco-2 cells. The Journal of Infectious Diseases. 1998;**178**(5):1373-1378

[83] Hirayama T et al. Helicobacter pylori vacuolating cytotoxin, VacA. Japanese Journal of Infectious Diseases. 2002;**55**(1):1-5

[84] Figura N et al. Cytotoxin production by *Campylobacter pylori* strains isolated from patients with peptic ulcers and from patients with chronic gastritis only. Journal of Clinical Microbiology. 1989;**27**(1):225-226

[85] Goossens H. Role of the vacuolating toxin from *Helicobacter pylori* in the pathogenesis of duodenal and gastric ulcer. Medical Microbiology Letters. 1992;**1**:153-159

[86] Rautelin H et al. Nonopsonic activation of neutrophils and cytotoxin production by *Helicobacter pylori*: Ulcerogenic markers. Scandinavian Journal of Gastroenterology. 1994;**29**(2):128-132

[87] Atherton J et al. Simple and accurate PCR-based system for typing vacuolating cytotoxin alleles of *Helicobacter pylori*. Journal of Clinical Microbiology. 1999;**37**(9):2979-2982

[88] Atherton J et al. Clinical and pathological importance of heterogeneity in vacA, the vacuolating cytotoxin gene of *Helicobacter pylori*. Gastroenterology. 1997;**112**(1):92-99

[89] Basso D et al. Analysis of *Helicobacter pylori* vacA and cagA genotypes and serum antibody profile in benign and malignant gastroduodenal diseases. Gut. 1998;**43**(2):182-186

[90] Evans DG et al. *Helicobacter pylori* cagA status and s and m alleles of vacA in isolates from individuals with a variety of H. pylori-associated gastric diseases. Journal of Clinical Microbiology. 1998;**36**(11):3435-3437

[91] Fallone CA et al. Association of *Helicobacter pylori* genotype with gastroesophageal reflux disease and other upper gastrointestinal diseases. The American Journal of Gastroenterology. 2000;**95**(3):659-669

Chapter 6

Helicobacter pylori Genes *jhp0940*, *jhp0945*, *jhp0947* and *jhp0949* are Associated with Gastroduodenal Disease

Romo-González Carolina and Coria-Jiménez Rafael

Additional information is available at the end of the chapter

http://dx.doi.org/10.5772/intechopen.81290

Abstract

The plasticity zone (PZ) of *Helicobacter pylori* is a genomic region harbouring genes that can be exchanged between strains, contributing to the genetic diversity of this bacterium. The presence or absence of genes reflects the adaptation and coevolution of a pathogen within its host. Among the genes present in PZ, *jhp0940, jhp0945, jhp0947* and *jhp0949* have gained attention due to their association with gastroduodenal disease, and the prevalence of the latter three among *H. pylori* isolates from different geographical regions has allowed this association to be understood. With respect to *jhp0940,* also known as *ctkA* (cellular translocation kinase A), various results have been obtained regarding its prevalence. However, the presence of *jhp0940* in isolates from children seems to be higher than that in isolates from adults, and the product of this gene can induce TNF-α, IL-6 and IL-8 via translocation of NF-κB into macrophages. While little is known about the functions of *jhp0945, jhp0947* and *jhp0949,* their presence in *H. pylori* strains induces IL-8 and IL-12 expression at higher levels than that in strains lacking these genes. In this chapter, we aim to show a general overview of the prevalence, association with gastroduodenal disease, and currently known function of the *H. pylori* genes *jhp0940, jhp0945, jhp0947* and *jhp0949,* which are located in PZ.

Keywords: plasticity zone, gastroduodenal diseases, *jhp0940, jhp0945, jhp0947, jhp094*

1. Introduction

Several techniques have been used to genotype *Helicobacter pylori* strains to identify differences at the genomic level between isolates from different populations as well as between isolates from the same individual, indicating the presence of a mixed infection [1–4]. Thus,

H. pylori is considered to have a high genetic diversity due to the presence or absence of genes, which indicates the adaptation and coevolution of the pathogen within different populations worldwide [5, 6]. Recently, advances in DNA sequencing technology have enabled the comparison of hundreds of sequences from the genomes of related bacteria. These comparative genome analyses between species have led to a concept that encompasses all of the genetic content within a bacterial species. This concept is referred to as a "pan-genome", which is defined as the complete genetic repertoire of a specific species, composed of both the central genome and the accessory genome, also known as the dispensable genome [7].

Comparative genomic studies of *H. pylori* began with the sequencing of strains J99 and 26,695. The *H. pylori* genome is approximately 1.6 million base pairs (1.6 Mb) in size, containing 1500 open reading frames (ORFs) and approximately 1500 protein-coding genes, with a G + C content of approximately 39%. Sequencing of these two strains showed that approximately 6–7% of the genes present in one strain are absent in the other. These genes are referred to as strain-specific genes, almost half of which are located in hypervariable regions within the *H. pylori* genome. Such regions are called plasticity zones (PZs), based on the variability of gene content between different isolates of *H. pylori*. These regions have a lower G + C content than the rest of the genome (34–35%), which suggests horizontal transfer [8]. The *cag* pathogenicity island (PAI) is a region exhibiting this reduced G + C content [9].

Several studies have shown that *H. pylori* strains gain and lose genes during an infection, which suggests that a continuous genetic flow occurs, primarily within the PZ [4, 6, 10, 11]. The frequency of the PZ genes, notably *jhp0940, jhp0945, jhp0947* and *jhp0949*, has been studied among *H. pylori* isolates from different geographical regions, which has demonstrated that some of these genes are associated with disease. Currently, some of these PZs have roles in *H. pylori* pathogenesis and are included in one of the three categories of *H. pylori* virulence classified by Yamaoka [12].

2. Genotyping of *H. pylori* strains

After strains J99 and 26695 were sequenced, the first DNA microarray could be designed [13], allowing the first genotyping studies of diverse strains of *H. pylori* to be conducted. The first study compared 15 *H. pylori* isolates and observed that 362 genes (22%) were variable among the strains, which were called strain-specific genes. Subsequently, another study explored the genomic diversity of *H. pylori* isolates from children and adults who presented a single or a mixed infection (the presence of more than one strain). The number of variable genes in individuals with a mixed infection was found to be higher than that in individuals infected with a single strain. No difference was observed in the number of strain-specific genes between strains from children and adults [14]. However, a study of 56 strains from different geographical regions observed that 25% of the genomic content between these strains corresponded to strain-specific genes [6]. The genetic content of the strains has also been correlated with their pathogenesis in an animal model [4, 15]. Moreover, Romo-González et al. [16] identified and compared the differences in the genetic content of strains isolated from Mexican patients with non-atrophic gastritis, duodenal ulcers and gastric cancer. The authors observed that the core genome shared by these isolates was composed of 1319 genes, while 341 genes (20.5%) were strain-specific among the assayed isolates, 37% of which were distributed within the PZs. The genotyping analysis provided an understanding of which genes are present and absent in the isolates associated with each disease, making it possible to associate some of these genes with specific diseases.

3. Structure of the *H. pylori* PZ

The PZ of strain J99 is the most studied and best characterised hypervariable region within the *H. pylori* genome. In the J99 *H. pylori* strain, the PZ is a continuous region from *jhp0916* to *jhp0961* [8]. Of the 48 ORFs that compose the PZ in strain J99, only six are present in strain 26695, and 10 are present in strain HPAG1. The PZs in strains J99 and 26695 are 45 and 68 kb in length, respectively. The PZ contains genes of unknown function as well as those encoding restriction-modification systems, topoisomerases, integrons, secretion systems, outer membrane proteins and transposons, the latter of which can be inserted into a genome without recombination. This insertion facilitates the propagation of transposons among bacterial species. These transposons are important because they can generate genomic deletions and rearrangements and even alter the expression of genes contiguous to their insertion site [17–19].

Kersulyte et al. [20] studied the nature of the PZ in the *H. pylori* genome by sequencing this locus in other strains and locating this area in strains that had been recently deposited in GenBank. The authors observed that the PZ is composed of inserted transposable elements

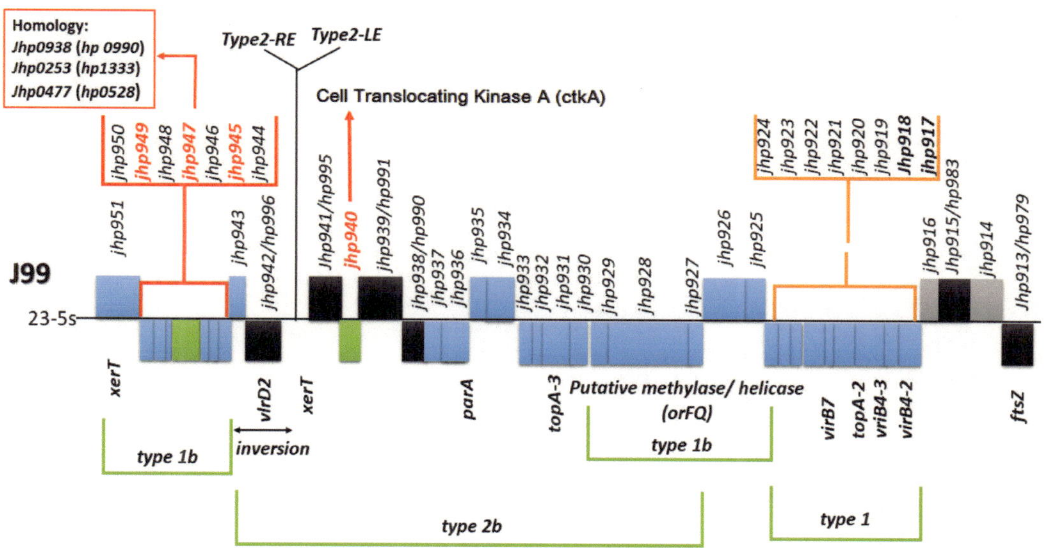

Figure 1. Plasticity zone of reference strain *Helicobacter pylori* J99 (*jhp0917-jhp0951*), figure adapted from [20]. This PZ is located between the *ftsZ* (*jhp0913*) gene and the 5S, 23S rRNA gene pair and is composed of TnPZ fragments: type 1 (*jhp0917-jhp0924*), type 1b (*jhp0944-jhp0951*; *jhp0925*, *jhp0926*, potentially part of *jhp0927-jhp0929* (*orfQ*) as well), and type 2 (*jhp0943-jhp0930*, potentially part of *orfQ* as well). Blue and green boxes indicate ORFs exclusive to J99 but not 26695 (another reference strain); black boxes indicate ORFs found in both strains; grey boxes indicate ORFs located between the *ftsZ* gene and the 5S, 23S rRNA gene pair that do not belong to the TnPZ.

that are flanked by discrete sequences of 5′AAGAATG and are each referred to as a TnPZ or "transposon, plasticity zone". Each TnPZ generally contains genes encoding type IV secretion proteins (tfs3), *xerT*, an ORF coding for a protein with helicase and DNA methylase domains and additional ORFs. Among the studied strains, several types of TnPZs with different gene arrangements or DNA sequence variations were observed and classified as type 1, type 1b and type 2 TnPZs. The genes *jhp0945*, *jhp0947* and *jhp0949* are located on a type 1b TnPZ, and *jhp0940* is located on a type 2 TnPZ in strain J99 (see **Figure 1**).

4. The PZ-associated gene *jhp0940* and its relationship with gastroduodenal disease

The prevalence of the *jhp0940* gene in various *H. pylori* isolates from different geographical regions has been explored (**Figure 2**). Studies have reported varying results regarding its prevalence and association with disease, and the presence of this gene has even been suggested to be related to a lower risk of peptic ulcers or gastric cancer [21, 22]. In a genotyping study of *H. pylori* isolates from Mexican individuals with gastroduodenal diseases, *jhp0940* was absent in all gastric cancer isolates [16]. In Brazil, in patients with duodenal ulcers or gastric cancer, the presence of *jhp0940* in *H. pylori* isolates exhibited no association with disease, as only 3 of 200 isolates had

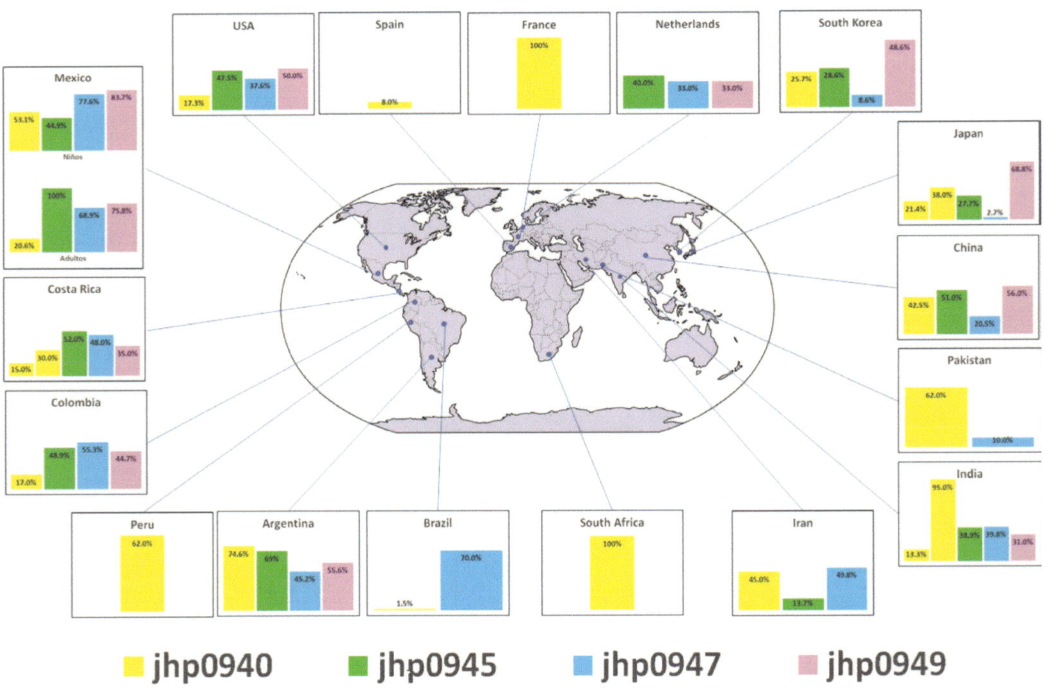

Figure 2. Frequency of the presence of the *jhp0940*, *jhp0945*, *jhp0947* and *jhp0949* genes among *H. pylori* isolates from different geographical regions. Frequencies are based on the following reports: USA, Colombia, South Korea [21]; Japan [21, 25]; Netherlands [32]; Iran [28]; India [22, 25]; Pakistan [27]; China [24]; Costa Rica [30, 25]; Brazil [23]; Argentina [26]; Peru [25]; South Africa [25]; France [25]; Spain [25]; and Mexico [16, 29].

the gene [23]. In addition, no association was found between this gene and disease in isolates from Chinese patients with chronic active gastritis, duodenal ulcers or gastric cancer [24].

A study in India observed less than a 20% prevalence of the *jhp0940* gene among *H. pylori* isolates [22]. However, in other isolates from India, the reported *jhp0940* prevalence was higher than 80% [25]. In Argentina, the frequency of the *jhp0940* gene was reported as 74.6%, and the frequency in isolates from individuals with chronic gastritis and peptic ulcer disease was similar, indicating no association with a specific disease [26]. However, Rizwan et al. [25] found no relationship between clinical results and the prevalence of *jhp0940* or the cag PAI in isolates from seven countries (India, South Africa, Japan, Costa Rica, Peru, France and Spain). The highest prevalence of this gene was observed in isolates from India, South Africa and France, in contrast to isolates from Spain, where less than 10% contained this gene.

However, Yakoob et al. [27] and Gholizade Tobnagh et al. [28] reported an association between *jhp0940* and gastric ulcers and gastric cancer, respectively. In general, the prevalence of this gene is lower than that of *jhp0945, jhp0947* and *jhp0949* in isolates from the different geographical regions that have been studied. There appears to be a difference between the presence of PZ genes between isolates from children and adults. Romo-González et al. [29] observed that the prevalence of *jhp0940* in *H. pylori* isolates from children is higher than that in adults and that the presence of the specific gene patterns (including *jhp0940-jhp0945-jhp0947-jhp0949* and *jhp0940-jhp0947-jhp0949*) is more common in isolates from children than in adults. Therefore, these authors suggest that this locus is more integrated in the early stages of infection, which could contribute to the bacterial virulence and evolution of the infection. These authors found no association between the presence of these four PZ-associated genes and the presence of *cagA*, cag PAI or *dupA*.

Romo-González et al. evaluated the *in vitro* expression of four PZ-associated genes in isolates from children and adults and observed the expression of only *jhp0945, jhp0947* and *jhp0949*, without significant differences among the expression of these three genes between the isolates obtained from children and adults. However, a correlation was observed among the expression of these three genes (unpublished data).

A possible explanation for the discordant results regarding the prevalence of *jhp0940* in different geographical regions is that geographical diversity exists in the sequence of this gene that does not allow for its identification with a single pair of primers. Another potential reason is that the absence of this gene in adult isolates is due to the loss of this gene during infection.

5. The PZ-associated genes *jhp0945, jhp0947* and *jhp0949* and their relationship with gastroduodenal disease

Studies on the prevalence of *jhp0945, jhp0947* and *jhp0949* among *H. pylori* isolates of different geographical origins have proposed that these genes are disease markers, suggesting that they could play a role in the pathogenesis of *H. pylori* [12]. This relationship was discovered in Costa Rica, where the prevalence of 21 ORFs present in the PZ of the J99 strain (*jhp0914-jhp0961*) was assessed in 17 strains from patients with gastric cancer and 26 strains from patients with gastritis. The results showed a high prevalence of *jhp0940* and *jhp0947* in patients with gastric cancer [30]. Later, in a study conducted in Brazil that included 200 *H. pylori* isolates from patients with duodenal ulcers, gastric cancer or gastritis, only *jhp0947* continued to show an association with gastric cancer and duodenal ulcers [23]. However, in another study of strains from Brazil, an association between *jhp0947* and peptic ulcers was not observed [31]. A different study assessed the prevalence of the *jhp0945-jhp947-jhp0949* locus in a Dutch population with gastritis and duodenal ulcers. In addition to *jhp0947*, the presence of *jhp0949* was associated with duodenal ulcers, whereas *jhp0945* was not associated with this disease [32]. Another cluster of genes for which the prevalence was examined consisted of *jhp0926*, *jhp0931*, *jhp0933*, *jhp0944* and *jhp0945* in isolates from Turkish patients with gastritis and peptic ulcers. Among these genes, *jhp0931* was the most prevalent in patients with peptic ulcers [33].

Yakoob et al. [27] observed that the *jhp0947* gene is more frequently present than *jhp0940* in strains associated with duodenal ulcers and gastric cancer. This association was determined to be independent of the presence of the virulence factor *cagA* in *H. pylori* strains. A study that included 296 Western isolates (from the United States and Colombia) and 217 East Asian isolates (from Korea and Japan) reported that the prevalence of *jhp0945*, *jhp0947* and *jhp0949* differs significantly between the two geographical regions. In the Western isolates, the presence of *jhp0945* was higher in isolates obtained from individuals with gastric ulcers, duodenal ulcers or gastric cancer than in those obtained from individuals with gastritis [21].

In *H. pylori* isolates of Chinese origin, the prevalence of *jhp0945*, *jhp0947* and *jhp0949* was significantly higher in individuals with duodenal ulcers and gastric cancer than in individuals with chronic gastritis [24]. Similarly, in isolates from India, the presence of *jhp0945*, *jhp0947* and *jhp0949* in *H. pylori* isolates was associated with disease [22]. The prevalence of *jhp0945*, *jhp0947* and *jhp0949* is associated with a greater risk of serious diseases in India [28]. PZ-associated genes (outside of locus *jhp0945-jhp0947-jhp0949*) that also have exhibited an association with disease include *jhp0950* and *jhp0917-jhp0918*. The *jhp0950* gene was associated with marginal zone B cell lymphomas (MZBL) and mucosa-associated lymphoid tissue (MALT) when its prevalence was examined in patients with gastritis, duodenal ulcers and gastric cancer [34]. Another gene considered a risk factor for the development of duodenal ulcers is the *dupA* gene (duodenal ulcer-promoting gene, *jhp0917-jhp0918*); its presence in strains from patients with this gastroduodenal pathology resulted in its association with this condition.

Predominant inflammation in the antrum region of the stomach as well as the infiltration of polymorphonuclear leukocytes can lead to the appearance of a duodenal ulcer [35]. The prevalence of the *dupA* gene among *H. pylori* strains varies according to the geographic region and duodenal pathology. The *dupA* gene is present in approximately 31% of strains in Asian countries and 64% of strains in Western countries [36–40].

However, two genotypes were observed for this gene, including strains with and without an extra 600 bp in the gene sequence [41]; another important characteristic is that it has high homology with VirB4, a component of the type IV secretion system (T4SS) of *H. pylori*, and recent studies suggest that *dupA* and the six homologues of adjacent vir genes (virB8-virB11, virD4 and virD2) in the PZ could form the third T4SS [42]. Many unanswered questions still exist regarding this gene and its role in *H. pylori* strains. To date, *jhp0947* is considered the best disease marker of the locus *jhp0945-jhp0947-jhp0949* because it seems to meet two of the conditions that Yamaoka et al. [12] suggest for an *H. pylori* virulence factor: (1) it has a disease or other *in vivo* correlation and (2) it is epidemiologically consistent across populations and regions.

6. Functional characteristics of the *jhp0940*, *jhp0945*, *jhp0947* and *jhp0949* genes

Among the *jhp0940*, *jhp0945*, *jhp0947* and jhp0949 genes, *jhp0940* is the best characterised. Rizwan et al. [25] purified recombinant JHP0940 protein, which, when incubated with macrophage cells (TPH-1), was able to induce the synthesis of TNF-α, IL-6 and IL-8 via translocation of NF-κB. This gene is currently known as *ctkA* (cell translocation kinase A), and the encoded protein has autophosphorylation activity [43]. In addition to being a serine/threonine kinase, CtkA can increase the phosphorylation of NF-κB by inducing TNF-α in a dose-dependent manner in HeLa cells [40]. However, Tenguria et al. [44] observed that *H. pylori* can secrete the CtkA and induce the expression of caspase-1 in macrophages (RAW264.7), generating an increase in the transcription of IL-1β and promoting the recruitment of infiltrated immune cells in the gastric mucosa. In addition, CtkA may be able to decrease cell viability through the Fas receptor. Recently, a study showed that CtkA is expressed from its native host and can induce stimulation of a pro-inflammatory response from gastric epithelial cells. This interaction is dependent upon a complement of the tfs3 T4SS genes but independent of the T4SS proteins encoded by either tfs4 or the *cag* PAI [45].

The function of the genes *jhp0945-jhp0947-jhp0949* is not yet well understood. However, it has been proposed that since they are consecutive genes and are oriented in the same direction, it is possible that they are expressed as an operon [30]. A study of the Dutch population revealed that the presence of these three genes in *H. pylori* strains induced higher amounts of IL-12 than in strain 26695, which does not possess these genes. However, the disruption of this locus reduces the production of IL-12 in THP-1 monocytic cells [32]. The presence of these three genes in strains from India induces a greater amount of IL-8 and induction of cell death by apoptosis (caspase-3 activity) in AGS cells than in strains that lack these genes [22]. *jhp0947* shows homology with *jhp0938* (*hp0990*) and *jhp0253* (*hp1333*), but their functions are still unknown. This gene also shows homology in the 5′ region with *jhp0477* (*hp0528*), which encodes a *virB9* homologue, an important component of the T4SS encoded by the *cag* PAI [46].

7. Conclusion

The PZ of the *H. pylori* genome contains several genes that have not been fully explored but could be important for understanding the pathogenesis of *H. pylori* due to their location in an area of the genome associated with genetic exchange. Among the most studied genes are *jhp0945*, *jhp0947* and *jhp0949*, which have been found to be associated with gastroduodenal disease, although their mechanism is still not clearly defined. However, the prevalence and association of *jhp0940* with ulcer or gastric cancer is still not entirely clear, although progress has been made in the characterisation of the function of this gene. It is important to continue exploring the presence and *in vivo* expression of these genes in strains isolated from children and adults from different geographical regions to elucidate their potential role in the pathogenesis of *H. pylori* infection.

Acknowledgements

We thank Mariana Espinosa and Miguel Angel Rojas for the help in generating the figures. The publication of this chapter was supported by funds from the National Institute of pediatrics, Mexico. No conflicts of interest exist for this paper.

Author details

Romo-González Carolina* and Coria-Jiménez Rafael

*Address all correspondence to: crgaro_06@yahoo.com.mx

Laboratory of Experimental Bacteriology, National Institute of Pediatrics,
Mexico City, Mexico

References

[1] Aviles-Jimenez F, Letley DP, Gonzalez-Valencia G, Salama N, Torres J, Atherton JC. Evolution of the *Helicobacter pylori* vacuolating cytotoxin in a human stomach. Journal of Bacteriology. 2004;**186**:5182-5185. DOI: 10.1128/JB.186.15.5182-5185.2004

[2] Camorlinga-Ponce M, Romo C, Gonzalez-Valencia G, Munoz O, Torres J. Topographical localisation of cagA positive and cagA negative *Helicobacter pylori* strains in the gastric mucosa; an in situ hybridization study. Journal of Clinical Pathology. 2004;**57**:822-828. DOI: 10.1136/jcp.2004.017087

[3] Ghose C, Perez-Perez GI, van Doorn LJ, Dominguez-Bello MG, Blaser MJ. High frequency of gastric colonization with multiple *Helicobacter pylori* strains in Venezuelan subjects. Journal of Clinical Microbiology. 2005;**43**:2635-2641. DOI: 10.1128/JCM.43.6.2635-2641.2005

[4] Israel DA, Salama N, Arnold CN, Moss SF, Ando T, Wirth HP, et al. *Helicobacter pylori* strain specifi differences in genetic content, identified by microrray, influence host inflammatory responses. The Journal of Clinical Investigation. 2001;**107**:611-620. DOI: 10.1172/JCI11450

[5] Mikkonen TP, Kärenlampi RI, Hänninen M. Phylogenetic analysis of gastric and enterohepatic *Helicobacter* species based on partial HSP60 gene sequences. International Journal of Systematic and Evolutionary Microbiology. 2004;**54**:753-758. DOI: 10.1099/ijs.0.02839-0

[6] Gressmann H, Linz B, Ghai R, Pleissner KP, Schlapbach R, Yamaoka Y, et al. Gain and loss of multiple genes during the evolution of *Helicobacter pylori*. PLoS Genetics. 2005;**1**:e43. DOI: 10.1371/journal.pgen.0010043

[7] Uchiyama I, Albritton J, Fukuyo M, Kojima KK, Yahara K, Kobayashi IA. Novel approach to *Helicobacter pylori* pan-genome analysis for identification of Genomic Islands. Cloeckaert A, ed. PLoS ONE. 2016;**11**(8):e0159419. DOI: 10.1371/journal.pone.0159419

[8] Alm RA, Ling LS, Moir DT, King BL, Brown ED, Doig PC, et al. Genomic-sequence comparison of two unrelated isolates of the human gastric pathogen *Helicobacter pylori*. Nature. 1999;**397**(6715):176-180. Erratum in: Nature 1999;**397**(6721):719. DOI: 10.1038/16495

[9] Censini S, Lange C, Xiang Z, Crabtree JE, Ghiara P, Borodovsky M, et al. cag a pathogenicity island of *Helicobacter pylori*, encodes type I-specific and disease-associated virulence factors. Proceedings of the National Academy of Sciences of the United States of America. 1996;**93**:14648-14653

[10] Dobrindt U, Hacker J. Whole genome plasticity in pathogenic bacteria. Current Opinion in Microbiology. 2001;**4**:550-557. DOI: 10.1016/S1369-5274(00)00250-2

[11] Janssen PJ, Audit B, Ouzounis CA. Strain-specific genes of *Helicobacter pylori*: Distribution, function and dynamics. Nucleic Acids Research. 2001;**29**:4395-4404

[12] Yamaoka Y. Roles of the plasticity regions of *Helicobacter pylori* in gastroduodenal pathogenesis. Journal of Medical Microbiology. 2008;**57**:545-553. DOI: 10.1099/jmm.0.2008/000570-0

[13] Salama N, Guillemin K, McDaniel TK, Sherlock G, Tompkins L, Falkow SA. Whole genome microarray reveals genetic diversity among *Helicobacter pylori* strains. Proceedings of the National Academy of Sciences of the United States of America. 2000;**97**:14668-14673. DOI: 10.1073/pnas.97.26.14668

[14] Salama NR, Gonzalez-Valencia G, Deatherage B, Aviles-Jimenez F, Atherton JC, Graham DY, et al. Genetic analysis of *Helicobacter pylori* strain populations colonizing the stomach at different times postinfection. Journal of Bacteriology. 2007;**189**:3834-3845. DOI: 10.1128/JB.01696-06

[15] Bjorkholm B, Lundin A, Sillen A, Guillemin K, Salama N, Rubio C, et al. Comparison of genetic divergence and fitness between two subclones of *Helicobacter pylori*. Infection and Immunity. 2001;**69**:7832-7838. DOI: 10.1128/IAI.69.12.7832-7838.2001

[16] Romo-González C, Salama NR, Burgeño-Ferreira J, Ponce-Castañeda V, Lazcano-Ponce E, Camorlinga-Ponce M, et al. Differences in genome content among *Helicobacter pylori* isolates from patients with gastritis, duodenal ulcer, or gastric cancer reveal novel disease-associated genes. Infection and Immunity. 2009;**77**:2201-2211. DOI: 10.1128/IAI.01284-08

[17] Kersulyte D, Velapatiño B, Mukhopadhyay AK, Cahuayme L, Bussalleu A, Combe J, et al. Cluster of type IV secretion genes in *Helicobacter pylori*'s plasticity zone. Journal of Bacteriology. 2003;**185**:3764-3772. DOI: 10.1128/JB.185.13.3764-3772.2003

[18] Fischer W, Breithaupt U, Kern BI, Smith SI, Spicher C, Haas R. A comprehensive analysis of *Helicobacter pylori* plasticity zones reveals that they are integrating conjugative elements with intermediate integration specificity. BMC Genomics. 2014;**2010**:15. DOI: 10.1186/1471-2164-15-310

[19] Dorer MS, Sessler TH, Salama NR. Recombination and DNA repair in *Helicobacter pylori*. Annual Review of Microbiology. 2011;**65**:329-348. DOI: 10.1146/annurev-micro-090110-102931

[20] Kersulyte D, Lee W, Subramaniam D, Anant S, Herrera P, Cabrera L, et al. *Helicobacter pylori*'s plasticity zones are novel transposable elements. PLoS One. 2009;**3**:e6859. DOI: 10.1371/journal.pone.0006859

[21] Sugimoto M, Watada M, Jung SW, Graham DY, Yamaoka Y. Role of *Helicobacter pylori* plasticity region genes in development of gastroduodenal diseases. Journal of Clinical Microbiology. 2012;**50**:441-448. DOI: 10.1128/JCM.00906-11

[22] Ganguly M, Sarkar S, Ghosh P, Sarkar A, Alam J, Karmakar BC, et al. *Helicobacter pylori* plasticity region genes are associated with the gastroduodenal diseases manifestation in India. Gut Pathogens. 2016;**8**:10. DOI: 10.1186/s13099-016-0093-5

[23] Santos A, Queiroz DM, Ménard A, Marais A, Rocha GA, Oliveira CA, et al. New pathogenicity marker found in the plasticity region of the *Helicobacter pylori* genome. Journal of Clinical Microbiology. 2003;**41**:1651-1655. DOI: 10.1128/JCM.41.4.1651-1655.2003

[24] Gong Y, Peng X, He L, Liang H, You Y, Zhang J. The distribution of jhp0940, jhp0945, jhp0947, jhp0949 and jhp0951 genes of *Helicobacter pylori* in China. BMC Gastroenterology. 2015;**15**:115. DOI: 10.1186/s12876-015-0341-z

[25] Rizwan M, Alvi A, Ahmed N. Novel protein antigen (JHP940) from the genomic plasticity region of *Helicobacter pylori* induces tumor necrosis factor alpha and interleukin-8 secretion by human macrophages. Journal of Bacteriology. 2008;**190**:1146-1151. DOI: 10.1128/JB.01309-07

[26] Armitano RI, Matteo MJ, Goldman C, Wonaga A, Viola LA, De Palma GZ, et al. *Helicobacter pylori* heterogeneity in patients with gastritis and peptic ulcer disease. Infection, Genetics and Evolution. 2013;**16**:377-385. DOI: 10.1016/j.meegid.2013.02.024

[27] Yakoob J, Abbas Z, Naz S, Islam M, Abid S, Jafri W. Associations between the plasticity region genes of *Helicobacter pylori* and gastroduodenal diseases in a high-prevalence area. Gut Liver. 2010;**4**:345-350. DOI: 10.5009/gnl.2010.4.3.345

[28] Gholizade Tobnagh S, Bakhti SZ, Latifi Navid S, Zahri S, Sadat Bakhti F. Role of plasticity region genes and cagE gene of cagPAI of *Helicobacter pylori* in development of gastrointestinal (GI) diseases. Asian Pacific Journal of Cancer Prevention. 2017;**18**:43-49. DOI: 10.22034/APJCP.2017.18.1.43

[29] Romo-González C, Consuelo-Sánchez A, Camorlinga-Ponce M, Velázquez-Guadarrama N, García-Zúñiga M, Burgueño-Ferreira J, et al. Plasticity region genes jhp0940, jhp0945, jhp0947, and jhp0949 of *Helicobacter pylori* in isolates from Mexican children. Helicobacter. 2015;**20**:231-237. DOI: 10.1111/hel.12194

[30] Occhialini A, Marais A, Alm R, Garcia F, Sierra R, Mégraud F. Distribution of open reading frames of plasticity region of strain J99 in *Helicobacter pylori* strains isolated from gastric carcinoma and gastritis patients in Costa Rica. Infection and Immunity. 2000;**68**:6240-6249. DOI: 10.1128/IAI.68.11.6240-6249.2000

[31] Proença Módena JL, Lopes Sales AI, Olszanski Acrani G, Russo R, Vilela Ribeiro MA, Fukuhara Y, et al. Association between *Helicobacter pylori* genotypes and gastric disorders in relation to the cag pathogenicity island. Diagnostic Microbiology and Infectious Disease. 2007;**59**:7-16. DOI: 10.1016/j.diagmicrobio.2007.03.019

[32] de Jonge R, Kuipers EJ, Langeveld SC, Loffeld RJ, Stoof J, van Vliet AH, et al. The *Helicobacter pylori* plasticity region locus jhp0947-jhp0949 is associated with duodenal ulcer disease and interleukin-12 production in monocyte cells. FEMS Immunology and Medical Microbiology. 2004;**41**:161-167. DOI: 10.1016/j.femsim.2004.03.003

[33] Salih BA, Abasiyanik MF, Ahmed N. A preliminary study on the genetic profile of cag pathogenicity-island and other virulent gene loci of *Helicobacter pylori* strains from Turkey. Infection, Genetics and Evolution. 2007;**7**:509-512. DOI: 10.1016/j.meegid.2007.03.002

[34] Lehours P, Dupouy S, Bergey B, Ruskoné-Foumestraux A, Delchier JC, Rad R, et al. Identification of a genetic marker of *Helicobacter pylori* strains involved in gastric extranodal marginal zone B cell lymphoma of the MALT-type. Gut. 2004;**53**:931-937. DOI: 10.1136/gut.2003.028811

[35] Lu H, Hsu PI, Graham DY, Yamaoka Y. Duodenal ulcer promoting gene of *Helicobacter pylori*. Gastroenterology. 2005;**128**:833-848. DOI: 10.1053/j.gastro.2005.01.009

[36] Arachchi HSJ, Kalra V, Lal B, Bhatia V, Baba CS, Chakravarthy S, et al. Prevalence of duodenal ulcer-promoting gene (dupA) of *Helicobacter pylori* in patients with duodenal ulcer in North Indian population. Helicobacter. 2007;**12**:591-597. DOI: 10.1111/j.1523-5378.2007.00557.x

[37] Argent RH, Burette A, Miendje Deyi VY, Atherton JC. The presence of dupA in *Helicobacter pylori* is not significantly associated with duodenal ulceration in Belgium, South Africa, China, or North America. Clinical Infectious Diseases. 2007;**45**:1204-1206. DOI: 10.1086/522177

[38] Douraghi M, Mohammadi M, Oghalaie A, Abdirad A, Mohagheghi MA, Eshagh Hosseini M, et al. dupA as a risk determinant in *Helicobacter pylori* infection. Journal of Medical Microbiology. 2008;**57**:554-562. DOI: 10.1099/jmm.0.47776-0

[39] Gomes LI, Rocha GA, Rocha AM, Soares TF, Oliveira CA, Bittencourt PF, et al. Lack of association between *Helicobacter pylori* infection with dupA-positive strains and gastroduodenal diseases in Brazilian patients. International Journal of Medical Microbiology. 2008;**298**:223-230. DOI: 10.1016/j.ijmm.2007.05.006

[40] Talebi Bezmin Abadi A, Perez-Perez G. Role of dupA in virulence of *Helicobacter pylori*. World Journal of Gastroenterology. 2016;**22**:10118-10123. DOI: 10.3748/wjg.v22.i46.10118

[41] Shiota S, Matsunari O, Watada M, Hanada K, Yamaoka Y. Systematic review and meta-analysis: The relationship between the *Helicobacter pylori* dupA gene and clinical outcomes. Gut Pathogens. 2010;**2**:13. DOI: 10.1186/1757-4749-2-13

[42] Jung SW, Sugimoto M, Shiota S, Graham DY, Yamaoka Y. The intact dupA cluster is a more reliable *Helicobacter pylori* virulence marker than dupA alone. Infection and Immunity. 2012;**80**:381-387. DOI: 10.1128/IAI.05472-11

[43] Kim DJ, Park KS, Kim JH, Yang SH, Yoon JY, Han BG, et al. *Helicobacter pylori* proinflammatory protein up-regulates NF-κB as a cell-translocating Ser/Thr kinase. Proceedings of the National Academy of Sciences of the United States of America. 2010;**107**:21418-21423. DOI: 10.1073/pnas.1010153107

[44] Tenguria S, Ansari SA, Khan N, Ranjan A, Devi S, Tegtmeyer N, et al. *Helicobacter pylori* cell translocating kinase (CtkA/JHP0940) is pro-apoptotic in mouse macrophages and acts as auto-phosphorylating tyrosine kinase. International Journal of Medical Microbiology. 2014;**304**:1066-1076. DOI: 10.1016/j.ijmm.2014.07.017

[45] Alandiyjany MN, Croxall NJ, Grove JI, Delahay RM. A role for the tfs3 ICE-encoded type IV secretion system in pro-inflammatory signalling by the *Helicobacter pylori* Ser/Thr kinase, CtkA. PLoS One. 2017;**12**:e0182144. DOI: 10.1371/journal.pone.0182144

[46] Tanaka J, Suzuki T, Mimuro H, Sasakawa C. Structural definition on the surface of *Helicobacter pylori* type IV secretion apparatus. Cellular Microbiology. 2003;**5**:395-404. DOI: 10.1046/j.1462-5822.2003.00286.x

Section 3

Helicobacter Pylori and Eradication Therapies

Chapter 7

Gastric Microbiota and Resistance to Antibiotics

Agnes Tving Stauning,
Rie Louise Møller Nordestgaard,
Tove Havnhøj Frandsen and Leif Percival Andersen

Additional information is available at the end of the chapter

http://dx.doi.org/10.5772/intechopen.80662

Abstract

Studies on gastric microbiota find several bacterial families and species in the stomach using molecular-based techniques. When biopsies are cultured, there may be growth of bacteria, pure culture of *Helicobacter pylori*, or no growth. When looking at the histological sections of corresponding biopsies no bacteria may be seen, except curved rods (*H. pylori*) adherent to the gastric epithelial cells. In a number of biopsies, several different bacteria are cultured with or without *H. pylori*. The non–*H. pylori* bacteria cultured are like the normal oral flora and may be contamination of the samples during endoscopy. In histological sections, these bacteria are seen above the mucin layer and not adherent to the epithelial cells confirming that it is contamination of the samples and can thus not be regarded as gastric microbiota. Therefore, the susceptibility of *H. pylori* to antibiotics is independent of coexisting bacterial flora. A review of *H. pylori* susceptibility to antibiotics in untreated and previous treated patients will be given including meta-analyses of *H. pylori* susceptibility to metronidazole (MTZ), clarithromycin, and levofloxacin. These data indicate that these antibiotics become more doubtful to use for primary therapy and should be banned for secondary therapy without susceptibility testing.

Keywords: gastric microbiota, *H. pylori*, histology, susceptibility testing, resistant rates

1. Introduction

Microbiota and microbiome are not always clearly defined or distinguished. The human microbiota comprises the population of microbial species that live on or in the human body. This is the resident flora of the body and does not include the transient flora (sampling contamination, etc.).

The microbiome is constituted by all the genes inside these microbial cells and is thus restricted to detection by molecular methods (sequencing, polymerase chain reactions [PCR]) [1].

By molecular methods, bacteria are usually identified to family and genera level [2]. Bacterial families and genera may include species and types of bacteria that may have completely opposite actions in the human body [3]. It is, therefore, doubtful if molecular methods alone are sensitive enough to predict the effect of the composition of microbiota. The limited original literature on gastric microbiota has mainly focused on gastric cancer and contains conflicting results [4–7]. There are many difficulties in investigating the gastric microbiota. One thing many authors are not aware of is the difficulty of getting samples without contaminating bacterial flora (**Figure 1**) [8]. In animal models, the whole stomach can be removed, and contamination of the stomach can be avoided, but in most animal species, physiology, acidity, etc. of the stomach are very different from the human stomach. Samples from the human stomach are usually taken as biopsies during gastroscopy. Even though the endoscope and the forceps are sterilized or decontaminated, it will be contaminated with oral bacterial flora during gastroscopy and thereby will the samples be contaminated by oral flora mainly of the phyla *Firmicutes* [8, 9].

Bacterial resistance to antibiotics can occur either if the bacteria obtain plasmids containing resistance genes from other bacteria in the microbiota (conjugation); they can take

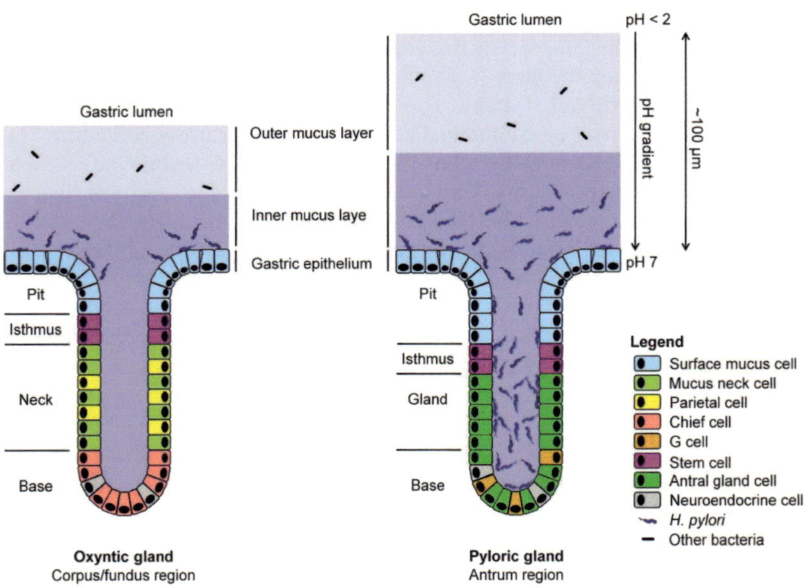

Figure 1. Schematic illustration of the gastric mucosa with the main cell types of oxyntic and pyloric glands in the gastric epithelium. Gastric stem cells reside in the isthmus zone of the gland and differentiate into precursors of the different cell lineages, which migrate either apically toward the gastric lumen or downwards to the base. The superficial epithelium and the gastric glands are covered by a viscous mucus layer mainly composed of MUC5AC, secreted by the SMCs, and MUC6, secreted mainly by MNCs and antral gland cells. The mucus layer consists of an inner layer, which is firmly attached to the epithelium, and an outer loose layer. The gastric pathogen *Helicobacter pylori* has been shown to use the transmucus pH gradient between the acidic gastric lumen and the near-neutral epithelial surface for spatial orientation to reach its niche at the juxtamucosal epithelium. The precise location of non–*H. pylori* microbiota is still hypothetical. [8].

up free DNA with resistance genes from the environment (transcription) or DNA can be transferred by bacteriophages (transduction). Furthermore, mutations can occur in the bacterial genome which may result in resistance if the mutation occurs in the part of the genome that codes for a structure on which the antibiotics act; this action may be interfered, and the bacteria becomes resistant to the antibiotic [10–12]. The conjugation of plasmids increases with the number of different bacteria in the microbiota and depends on a close contact between the bacteria. Uptake of free DNA does not demand a direct contact with other bacteria, but bacteria should probably be present in the close environment [3]. Mutations occur in all bacteria with a certain time because of natural replication errors [12]. Some bacteria mutate more often than others; but because of the short generation time for bacteria, each bacterial clone will have several mutations. If the mutation occurs in a part of the genome, which is target for the antibiotics, resistance to the antibiotic may occur.

2. Study on gastric microbiota

In a previous unpublished study that included 411 biopsies from patients undergoing upper gastrointestinal endoscopy were investigated both by microaerobic culture and by histology (**Table 1**). From 249 (60%) biopsies other bacteria than *H. pylori* were cultured. These bacteria were oral flora, that is, *Streptococcus* spp., *Staphylococcus* spp., *Corynebacterium* spp., *Neisseria* spp., etc., which may indicate contamination of both the endoscope and the biopsies during the procedure. In histological sections, very few bacteria except *H. pylori* were seen in 20 (5%) of the biopsies. In all cases, the bacteria were located superficial to the mucus layer and not in relation to the epithelial cells and *H. pylori*, which confirm that it is contamination from the oral cavity. The discrepancy in the number of biopsies with other bacteria than *H. pylori* between culture and histology may be because very few bacteria (less than 5 colonies) are cultured and the preparation of histological sections may remove much of the mucin and the contaminating bacteria. *H. pylori* was found alone without contamination in 60 biopsies by culture and in 83 biopsies by histology which indicate that *H. pylori* is a true gastric microbiota (**Figure 2**).

All known mechanisms for *H. pylori* resistance to all antibiotics are point mutations located on the chromosome (**Table 2**), indicating no uptake of plasmids or free DNA, which support that *H. pylori* is the only bacteria in the true gastric microbiota and everything else is transient contaminating flora [13].

No. of biopsies	**Culture**		**Histology**	
	H. pylori	**Other bacteria**	***H. pylori***	**Other bacteria**
411	106	249	83	20

Table 1. Comparison of culture and histological finding of *H. pylori* and other bacteria (oral flora) in gastric biopsies.

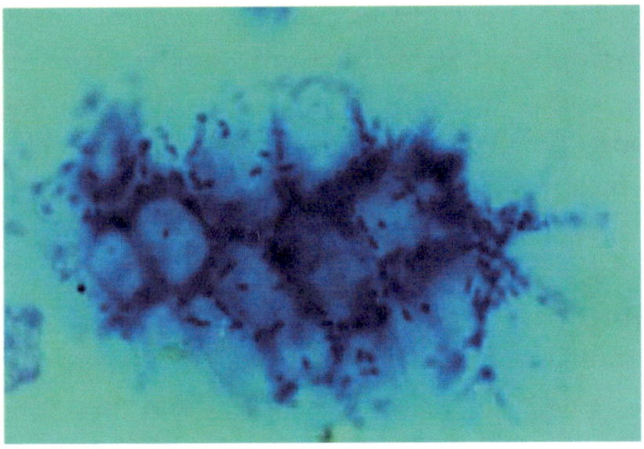

Figure 2. Imprint cytology showing the presence of *H. pylori* (Giemsa stain, ×400) Rahbar [84].

Resistance to	Mutation
Amoxicillin	PBP1
Clarithromycin	InfB
	rp1V
	A2142C
	A2142G
	A2143G
Metronidazole	rdxA
	frxA
	fdxB
Fluoroquinolones	gyrA
	gyrB
Tetracycline	AGA925-967TTC
Rifampicin	RNA polymerase subunit beta/beta

Table 2. Examples of mutations in *H. pylori* causing resistance to antibiotics.

3. Diagnosis of *H. pylori*

The detection of *H. pylori* can be done by invasive and noninvasive methods. The invasive methods require a biopsy, whereas the noninvasive methods are gentler for the patient.

Culture of *H. pylori* may be difficult and the sensitivity may be rather low (50–85%) [14]. The sensitivity of the culture depends on transport time to the lab and the culture method used [15]. Different agar plates or incubation time can also give different results on the same biopsy. Two biopsies from the antrum and two biopsies from the fundus are preferred when making a culture as *H. pylori* is unevenly distributed in the stomach. Culture is the only method by which it is possible to make a full susceptibility test.

Histology is an invasive method which requires a least one antral biopsy and preferably two antral and two corpus biopsies. The biopsy is stained with hematoxylin and eosin, Giemsa, or silver staining. *H. pylori* is identified by the color, shape, and close relation to the mucosa and can be confirmed by immunohistochemistry using *H. pylori*–specific antibodies. The histology has shown to have a sensitivity at the same level as culture but is influenced by the size of the biopsy [14]. The number of biopsies and the location in the stomach also modify the sensitivity. The specificity of histology is lower than the specificity of the culture as histology cannot distinguish *H. pylori* from non–pylori *Helicobacter* species. The detection rates in cultures and histology varies with varying expertise of examiners. If the patient is taking proton pump inhibitor (PPI), bismuth, or antibiotics prior to gastroscopy, it might change the shape of *H. pylori* from curved rod to a coccoid form. This form is undetectable in the routine microscopy technique and requires fluorescent *in situ* hybridization, immunohistochemistry with specific antibodies to *H. pylori*, or confirmation by the 16s rRNA and 23rRAN sequencing, which are irrespective of the shape of the bacteria [16].

H. pylori urease breaks down urea to ammonia and carbon dioxide. This feature is used in the diagnostic methods "rapid urease test" (RUT) and "urea breath test" (UBT). RUT is an invasive method that preferably needs two biopsies. If the biopsy contains *H. pylori*, the release of ammonia increases the pH of the test medium, which is seen by a color change due to a pH indicator. The result of the test is fast and takes approximately ½ hour. UBT is a noninvasive method where the patient ingests 13C-labeled urea. If the patient is infected with *H. pylori*, orally ingested 13C-urea is broken down to 13C-labeled carbon dioxide, which is then exhaled. The sensitivity of the two tests is 75–85% for RUT and >95% for UBT. Likewise, the UBT has a higher specificity (<95%) when compared to RUT (85–95%). For both RUT and UBT, PPI and antibiotics can give false negative results. Furthermore, coccoid forms of *H. pylori* would not produce urease and would therefore give a false negative result [17].

Stool antigen test is another noninvasive method. It was first successfully described in 1997 using polyclonal antibodies [18]. Today monoclonal antibodies are used, and the sensitivity and the specificity are at the same levels as for UBT, but are preferred in special patients like children and patients with bleeding ulcers. This test can be done within ½ hour and is good for screening a patient for an infection with *H. pylori*. Despite this, antigen excretion may vary over time, and antigens may degrade while passing through the intestines, which may lead to false negative results.

The humoral antibody response to *H. pylori* can be measured by either serum IgG antibodies to *H. pylori*, which shows an ongoing or a previous infection, or by serum IgM antibodies, which shows an ongoing acute infection. *H. pylori* IgG antibodies can be detected in sputum or urine but have a much lover sensitivity and specificity than serum antibodies. Antibodies to *H. pylori* in serum can be tested by ELISA or "near patient test (NPT)." NPT uses immunechromatography or passive agglutination. A 2013 study compared the NPT and the ELISA test. The study showed that the NPT never reach 90% in sensitivity, and the frequency of false negatives and false positives were high [19]. Several tested ELISA kits showed a high specificity and sensitivity above 90%. However, the serological kits may differ considerably depending on the antigens that are included in the kit as antibodies to low-molecular-weight antigens (outer membrane antigens) decline significantly within 3 months, whereas antibodies to high-molecular-weight antigens (CagA, VacA, etc.) may stay potent for years [20]. CagA antibodies remain stable for a long period of time and can probably be useful for the detection of *H. pylori* infections in patients with gastric cancer when other tests are negative [21]. Due to local strain distribution of *H. pylori,* the serology kits should be made by using local *H. pylori* strains, and the kits should be locally validated [21].

Gastrin and pepsinogen are compounds produced in the stomach that depend on the changes in the gastric mucosa, and the serum levels of pepsinogens are a marker of atrophic gastritis [22]. This can be combined with the *H. pylori* antibody test to predict the risk of developing gastric cancer.

Molecular methods have been of increasing interest in the field of microbiology and for detection of *H. pylori*. Polymerase chain reaction (PCR) seems to be more sensitive than any other method to detect *H. pylori* [23]. The main problem is that the method does not distinguish between live bacteria and DNA from dead bacteria. Real-time PCR (RT-PCR), which is a fast and quantitative PCR, seems to be more sensitive than classical PCR [24]. By sequencing the 16S RNA or 23S RNA region, it is possible to detect *Helicobacter* species and susceptibility to clarithromycin and tetracycline [25–27]. However, it is a more expensive and time-consuming method. A commercial kit has combined detection of *H. pylori* and susceptibility to clarithromycin in a classical PCR. However, culture is still needed for a full susceptibility testing. There are so many point mutations causing resistance to antibiotics in *H. pylori* that a full susceptibility analysis can only be detected by whole genome sequencing [28].

4. *H. pylori* susceptibility to antibiotics

During the last decade, an increased number of *H. pylori* have become resistant to antibiotics, especially to clarithromycin and levofloxacin [29]. The resistance rates to metronidazole have always been more than 15% worldwide, but the increasing resistance rates to clarithromycin and levofloxacin in some areas have become higher than 10–15%. Thus, these antibiotics are not recommended for first-line therapy of *H. pylori* without prior susceptibility testing [21]. It is common to treat *H. pylori* infections without prior susceptibility testing, and different studies show a much lower resistance rate to clarithromycin in *H. pylori* from untreated patients than in *H. pylori* from previously treated patients [30–32]. It is therefore of the greatest importance to make susceptibility testing after the first treatment failure.

The susceptibility testing of *H. pylori* can be done by various methods. The most common are dilution methods, disk diffusion, and E-test.

The dilution method is regarded to be the golden standard for susceptibility testing. A two-fold dilution row of the test antibiotic is made. A standard number of bacteria (McFarland 3) are added to each tube with antibiotics. The bacterial growth is inhibited by high concentrations of antibiotics. The first tube with bacterial growth is called the minimal inhibitory concentration (MIC). *H. pylori* should be grown for 48–72 hours under microaerobic conditions. It may be difficult to find a suitable media in which *H. pylori* grows fast enough, and the slightest contamination will grow faster than *H. pylori* and thereby spoil the susceptibility testing.

The disk diffusion test requires a small tablet of an antibiotic. The tablet is placed on the agar plate and is incubated for 3 days. After 3 days, there will be a zone around the tablet with no

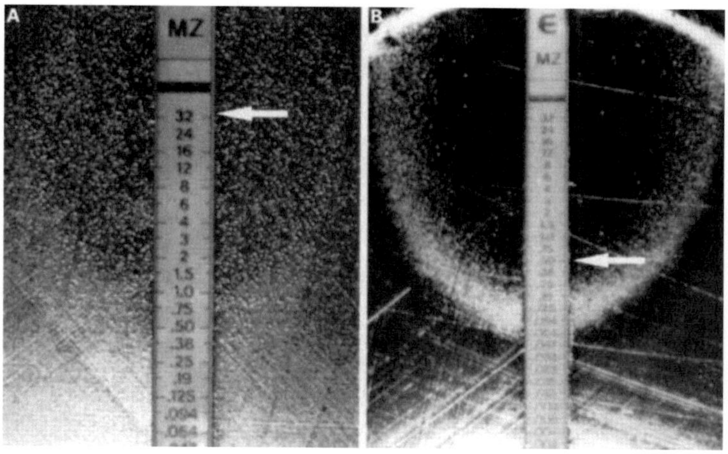

Figure 3. Reading guide for E tests. (A) Colonies of a metronidazole-resistant subpopulation in the ellipse minimum inhibitory concentration (MIC) >32; (B) trailing of microcolonies at the end point MIC 0.5 μg/ml. Warburton-Timms and McNulty [85].

growth of *H. pylori*. This is the inhibition zone, and the diameter of the zone can be translated to an MIC value, which shows whether or not the bacteria are resistant to the antibiotic. To make the susceptibility testing of *H. pylori*, a McFarland 3.0 dilution of *H. pylori* and Mueller-Hinton agar plates with 10% blood or chocolate ager plates should be used and incubated in microaerobic conditions at 37°C.

The E-test is a stripe with a concentration gradient of an antibiotic. The stripe is placed on the agar plate and is incubated for 3 days. After 3 days, there will be a droplet shape around the stripe with no growth of *H. pylori* (**Figure 3**). That concentration where *H. pylori* grows close to stripe is the MIC value [33].

5. Treatment of *H. pylori* infection

H. pylori infections are usually treated with a combination of antibiotics and nonantibiotics (proton pump inhibitor [PPI] or bismuth salts). Usually, a combination of two or three antibiotics is used, as the effect of monotherapy has been found insufficient. The most commonly used antibiotics are amoxicillin, clarithromycin, metronidazole, fluoroquinolones, tetracycline, and rifampicin (**Table 3**).

H. pylori is found in very different environments such as the gastric lumen with a relatively low pH, in between the epithelial cells and on the basement membrane with a neutral pH but protected as intracellular microorganisms. When choosing antibiotics, it is important to select antibiotic to which *H. pylori* is sensitive and is active in all the environmental niches where *H. pylori* occurs. It is also important to look at the duration of the efficacy of antibiotics to keep stable levels above the minimal inhibitory concentrations.

PPI in standard doses do not have antibacterial effect on *H. pylori*, but 5–10 times higher doses have a direct effect on *H. pylori*. Bismuth salts binds to the surface of *H. pylori* but have

Group	Preparation
Antibiotics	Amoxicillin
	Clarithromycin
	Metronidazole
	Tetracycline
	Levofloxacin
	Ciprofloxacin
	Rifampicin
Nonantibiotics	PPI
	Bismuth nitrate
	Bismuth citrate
	Bismuth subsalicylate
	H_2 blocker

Table 3. Commonly used antibiotics and nonantibiotics for treatment of *H. pylori* infections.

a relatively little antibacterial effect. However, bismuth salts affect the respiratory chain at the same points as metronidazole and thereby reverts metronidazole resistance in *H. pylori* and thus becomes sensitive to metronidazole.

6. Prevalence of *H. pylori* resistance to antibiotics

When analyzing different studies around the world, the primary resistance rate for *H. pylori* varies. The highest rate of primary metronidazole (MTZ) resistance is found in Africa (52%) followed by South America (49%) and Asia (43%). The lowest resistance rate is found in Europe (35%). The highest primary resistance rates for clarithromycin and levofloxacin are found in South America (20 and 27%) while the lowest rates are found in Europe (12 and 10%) [30–32, 34–67]. There is a significantly ($p < 0.001$) higher risk of primary metronidazole and levofloxacin resistance in Asian when compared to Europe.

The high rate of metronidazole resistance seen in developing countries may be due to the high use of metronidazole for treatment of parasites and gynecological infections [62, 68]. It is therefore likely that the patients who are treated for *H. pylori* with metronidazole for the first time are resistant for this treatment. It is recommended to use bismuth therapy together with metronidazole in the first-line treatment in areas with high metronidazole resistance [21].

The high resistance rates for clarithromycin and levofloxacin in South America, Africa, and Asia can be due to the use of huge amounts of antibiotics in general [69]. Typically, the diagnostics are not precise, and the patients are treated with more a broad spectrum of antibiotics for a longer period. This can lead to a faster development of resistance in *H. pylori* [70].

A large multinational study tested *H. pylori* resistance in 18 European countries [29]. All 18 countries used E-test for the susceptibility testing and only tested patients who had never been treated for *H. pylori* before. In total, 2204 people were included in the study, and the resistance rate for adults were 18% for clarithromycin, 14% for levofloxacin, and 35% for metronidazole. They found a significant association between the use of only long-acting macrolides and clarithromycin resistance. The levofloxacin resistance was significantly associated with the use of quinolone.

The prevalence of *H. pylori* resistance to antibiotics was tested in Denmark in 1997, 1998–2004, and 2013 [71–73]. Throughout the years, the resistance for clarithromycin has increased from 0% in 1997 to 53% in 2013, and likewise, the resistance for metronidazole increased from 20 to 74% [12–14]. None of the studies mention whether or not the patients have had *H. pylori* eradication therapy prior to testing or not, which might explain the huge increase in resistance.

6.1. Effect of antibiotic treatment on *H. pylori* resistance rates

International guidelines recommend first line of treatment of *H. pylori* infections with 10 days of triple therapy (PPI, clarithromycin, and metronidazole or amoxicillin). If this fails, a treatment with four types of medicine (PPI, bismuth subsalicylate, tetracycline, and metronidazole) for 2 weeks is recommended. After treatment failure for the second time, it is recommended to perform a gastroscopy and susceptibility testing for *H. pylori* [21].

The primary and secondary resistance rate for *H. pylori* has only been described in eight studies [30, 32, 40, 43, 58, 65, 66, 74]. By using "Review Manager 5.3," it is possible to compare the studies via Forest plots. The meta-analyses show that the secondary resistance is significantly higher ($p < 0.001$) than the primary.

The meta-analysis shows a high increasing resistance rate for all three antibiotics when the patient had been treated for *H. pylori* previously. The high and increasing resistance rates to metronidazole, clarithromycin, and levofloxacin make it uncertain that these antibiotics should be recommended as the first-line therapy of *H. pylori* infections without prior endoscopy and susceptibility testing (**Figure 4A–C**).

6.2. Vaccine

Another way to overcome *H. pylori* infections is with a vaccine. In the past couple of years, many studies have investigated developing an effective and safe vaccine. The development of an effective vaccine is complicated by the noninvasive nature of *H. pylori.* It stays in the lumen of the stomach and does not cross the epithelium. Therefore, the vaccine should affect T helper memory cells, which are required to stay in the lumen during a *H. pylori* infection [75].

Appropriate bacterial antigens, safe and effective adjuvants, and a route of delivery are required for developing a vaccine. For the bacterial antigen, most studies use urease, but other antigens are investigated for example Cag L. The CagL is a protein essential for the pathogenesis of *H. pylori.* It binds to integrins in the mucosa and triggers the release of the carcinogen CagA to the host cells

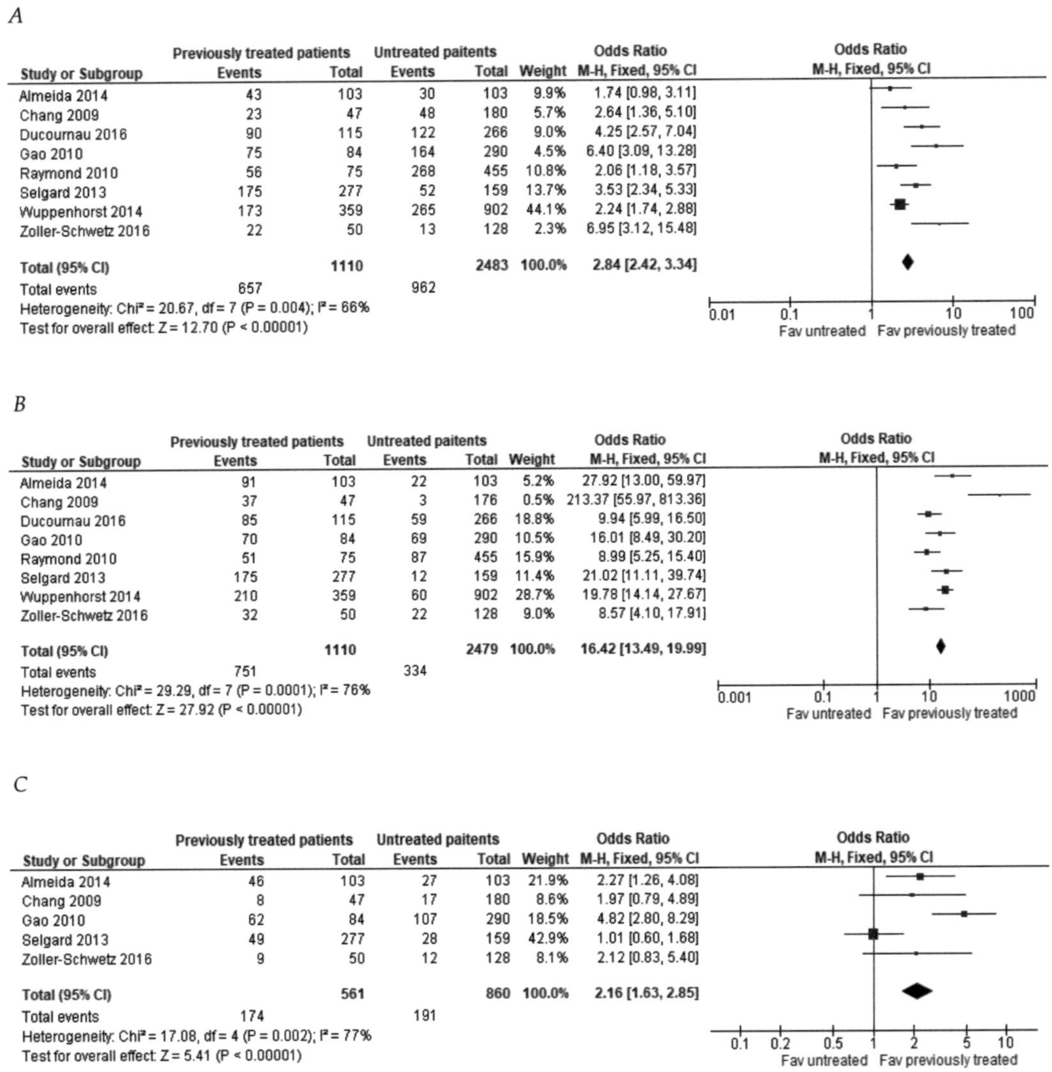

Figure 4. Meta-analysis for MTZ (A), CLR (B), and LEV (C). For all three antibiotics, there is a higher odds ratio for resistance if the patient is previously treated for infection with *H. pylori*.

through the type IV secretin system. CagL also introduces an IL-8 response, which causes inflammation [76]. The use of CagL in a subunit vaccine was investigated by Choudhari et al. in 2013 [75]. The study showed that CagL was stable in pH 4–6 and that sucrose enhances the stability.

The use of heat shock proteins in a vaccine introduced protective immunity without requiring the addition of an adjuvant. The protection, however, is not optimal because sterilizing immunity is not obtained, which is shown in a study from 2014 [77].

A derivate of the cholera toxin (CTA1-DD) and safe nontoxic mutants of *Escherichia coli* heat labile toxin (dm2T) have also been tested as potential adjuvants. CTA1-DD enhances the Th1 and Th17 immunity and reduces the bacterial colonization by three- to eight-fold [78]. The use of dm2T was equally as effective as the gold standard *H. pylori* vaccine containing cholera toxin [79].

The routes of delivery that have been tested are sublingual, intranasal, respiratory, and oral [79]. A study on humans from China (2015) tested a vaccine based on a urease B subunit and heat-labile enterotoxin B subunit (gene derived from *E. coli* H44815) [80]. The vaccine was taken orally three times (day 0, 14, and 28). This study showed a vaccine efficacy of 71.8% in the first year, 55% in in the second year, and 55.8% in the third year after vaccinations. Even though these findings are excellent, a 100% effective vaccine is still not developed. More studies and longer time follow-ups are needed before a fully effective vaccine is on the market. If a fully effective vaccine is made, it would be the best heath measure against *H. pylori* infections and gastric cancer.

7. Discussion

The human gastric microbiota may be difficult to estimate since samples for microbiome investigations often are contaminated with oral bacterial flora during gastroscopy. And the studies in these fields do not make any attempt to remove the oral contamination prior to sequencing. Histological examination of biopsies reveals *H. pylori* as the only bacteria in close relation to the epithelial cells in the gastric mucosa. When *H. pylori* is seen in stomach samples, there is always a strong humoral and cellular immune response to *H. pylori* and it thereby fulfills the criteria for a true infection but also a colonization. This has not been shown for any other bacteria.

Thus, in noncancer patients, *H. pylori* seems to be the gastric microbiota. In patients with gastric cancer, there may be a different situation as the mucosa is disintegrated and an overgrowth of intestinal bacteria is common. However, it remains to be shown that the intestinal bacteria adhere to the gastric mucosa and cause a local immune response. It is, therefore, believed that *H. pylori* is still the most important gastric pathogen.

An increasing resistance to antibiotics in *H. pylori* has been seen worldwide especially to metronidazole, clarithromycin, and levofloxacin. This is a worrying development as it may interfere with our recommendations for primary treatment of *H. pylori* without susceptibility testing. It is a question how fast the resistance occurs. Should susceptibility testing be done after first treatment failure or can it wait until the second treatment failure as recommended? At least the resistance rates are much higher in previously treated patients than in untreated patients.

Due to the high resistant rates, it is necessary to perform a susceptibility test before starting the treatment. The advantages would be a better and maybe quicker eradication of the *H. pylori* infection. Disadvantages of early susceptibility testing are the cost and time of the analyses. Biopsies are an invasive method and may often be painful for the patient. Furthermore, it takes up to 14 days before a full susceptibility test is completed, so the real treatment starts approximately 2 weeks after the doctor confirms the presence of *H. pylori*. By this time, the patient could have been done with the first line of treatment. In the short perspective, a quick susceptibility test would be very time consuming, but in the long perspective, it might save the patient from several treatments and prevent the relapse of the *H. pylori* infection. But it also gives a better overview on how quickly *H. pylori* develops resistance to the recommended treatment.

When detecting *H. pylori,* the best would be a quick a method that was as quick as PCR but also made it possible to have a full susceptibility test incorporated. New primers for detecting antibiotic resistance are in progress, but the problem is that there are many different mutations leading to the same resistance profile. *H. pylori* only develops antibiotic resistance by mutation in the genome. For MTZ, mutations in at least nine different genes are known to contribute to MTZ resistance [13]. If the detecting of MTZ resistance should be made by PCR, it would be necessary to perform the PCR with many different primers all looking for one specific mutation. In theory, this would be the most sensitive way to find MTZ resistance, but in practice, it would be almost impossible, take a lot of time, and would be expensive.

Due to the enormous amount of mutations leading to antibiotic resistant, the culture and susceptibility testing done by E-test is still the best and most economical way.

The increasing resistant rates to the most commonly used antibiotics raises the question of whether other antibiotics or combinations of antibiotic and nonantibiotic should be used for primary treatment of *H. pylori* infections without susceptibility testing. Bismuth compounds in standard doses, proton pump inhibitors, and acid suppressing compounds in high doses may convert the MTZ resistance [81]. This makes MTZ useful in combination with these compounds, especially the bismuth compounds, which have been shown in clinical studies [21]. Nonantibiotics such as neuroleptics and other compounds acting on the central nerves system have anti–*H. pylori* effect *in vitro* [82] and compounds without effect on the central nervous system may be candidates for alternative treatment. Herbs like broccoli and green tee have some effect on *H. pylori* and may in combination with antibiotics and nonantibiotics be candidates for treatment in the future [83].

8. Conclusion

H. pylori is the most important gastric pathogen and may constitute the true gastric microbiota. It is, therefore, important to follow the development of resistance in *H. pylori* to antibiotics. With the increased resistance of *H. pylori* to metronidazole, clarithromycin, and levofloxacin, it may be doubtful if these antibiotics can be recommended as primary treatment without susceptibility testing.

Conflict of interest

The authors declare no conflicts of interests.

Author details

Agnes Tving Stauning, Rie Louise Møller Nordestgaard, Tove Havnhøj Frandsen
and Leif Percival Andersen*

*Address all correspondence to: leif.percival.andersen@regionh.dk

Department of Clinical Microbiology, The Helicobacter Research Center, Copenhagen
University Hospital (Rigshospitalet), Copenhagen, Denmark

References

[1] Ursell LK, Metcalf JL, Parfrey LW, et al. Defining the human microbiome. Nutrition Reviews. 2012;**70**(Suppl 1):S38-S44

[2] Jackson MA, Bonder MJ, Kuncheva Z, et al. Detection of stable community structures within gut microbiota co-occurrence networks from different human populations. PeerJ. 2018;**6**

[3] Jorgensen JH, Pfaller MA, Karen C, et al. Manual of Clinical Microbiology. 11th ed. Washington DC: ASM Press; 2015

[4] Aviles-Jimenez F, Vazquez-Jimenez F, Medrano-Guzman R, et al. Stomach microbiota composition varies between patients with non-atrophic gastritis and patients with intestinal type of gastric cancer. Scientific Reports. 2015;**4**(1):4202

[5] Wang L, Zhou J, Xin Y, Geng C, et al. Bacterial overgrowth and diversification of microbiota in gastric cancer. European Journal of Gastroenterology & Hepatology. 2016;**28**(3):261-266

[6] Bik EM, Eckburg PB, Gill SR, et al. Molecular analysis of the bacterial microbiota in the human stomach. Proceedings of the National Academy of Sciences. 2006;**103**(3):732-737

[7] Maldonado-Contreras A, Goldfarb KC, Godoy-Vitorino F, et al. Structure of the human gastric bacterial community in relation to *Helicobacter pylori* status. The ISME Journal. 2011;**5**(4):574-579

[8] Yang I, Nell S, Suerbaum S. Survival in hostile territory: the microbiota of the stomach. FEMS Microbiology Reviews. 2013;**37**(5):736-761

[9] Liu X, Nie W, Liang J, et al. Interaction of *Helicobacter Pylori* with other microbiota species in the development of gastric cancer. Archives of Clinical Microbiology. 2017;**8**(2)

[10] Carroll AC, Wong A. Plasmid persistence: Costs, benefits and the plasmid paradox. Canadian Journal of Microbiology. May 2018;**64**(5):293-304. cjm-2017-0609

[11] Dorward DW, Garon CF. DNA-binding proteins in cells and membrane blebs of *Neisseria gonorrhoeae*. Journal of Bacteriology. 1989;**171**(8):4196-4201

[12] Durão P, Balbontín R, Gordo I. Evolutionary mechanisms shaping the maintenance of antibiotic resistance. Trends in Microbiology. Aug 2018;**26**(8):677-691

[13] Arslan N, Yılmaz Ö, Demiray-Gürbüz E. Importance of antimicrobial susceptibility testing for the management of eradication in *Helicobacter pylori* infection. World Journal of Gastroenterology. 2017;**23**(16):2854

[14] Bytzer P, Dahlerup JF, Eriksen JR, et al. Diagnosis and treatment of *Helicobacter pylori* infection. Danish Medical Bulletin. 2011;**58**(4):1-5

[15] Cuchi E, Forné M, Quintana S. Comparison of two transport media and three culture media for primary isolation of *Helicobacter pylori* from gastric biopsies. European Society of Clinical Microbiology and Infectious Diseases. 2002;**8**:609-610

[16] Patel SK, Pratap CB, Jain AK, et al. Diagnosis of *Helicobacter pylori:* What should be the gold standard? World Journal of Gastroenterology. 2014;**20**(36):12847-12859

[17] Koletzko S. Noninvasive diagnostic tests for *Helicobacter pylori* infection in children. Canadian Journal of Gastroenterology. 2005;**19**(7):433-439

[18] Makristathis A, Pasching E, Schütze K, et al. Detection of *Helicobacter pylori* in stool specimens by PCR and antigen enzyme immunoassay. Journal of Clinical Microbiology. 1998; **36**(9):2772-2774

[19] Burucoa C, Delchier JC, Courillon-Mallet A, et al. Comparative evaluation of 29 commercial *Helicobacter pylori* serological kits. Helicobacter. 2013;**18**(3):169-179

[20] Andersen LP, Espersen F, Souckova A, et al. Isolation and preliminary evaluation of a low-molecular-mass antigen preparation for improved detection of *Helicobacter pylori* immunoglobulin G antibodies. Clinical and Diagnostic Laboratory Immunology. 1995;**2**(2):156-159

[21] Malfertheiner P, Megraud F, O'morain CA, et al. Management of *Helicobacter pylori* infection—the Maastricht V/florence consensus report. Gut. 2017;**66**:6-30

[22] Shimoyama T, Oyama T, Matsuzaka M, et al. Comparison of a stool antigen test and serology for the diagnosis of *Helicobacter pylori* infection in mass survey. Helicobacter. 2009; **14**(2):87-90

[23] Cosgun Y, Yildirim A, Yucel M, et al. Evaluation of invasive and noninvasive methods for the diagnosis of *Helicobacter Pylori* infection. Asian Pacific Journal of Cancer Prevention. 2016;**17**(12):5265-5272

[24] Monno R, Giorgio F, Carmine P, et al. *Helicobacter pylori* clarithromycin resistance detected by Etest and TaqMan real-time polymerase chain reaction: A comparative study. APMIS. 2012;**120**(9):712-717

[25] Redondo JJ, Keller PM, Zbinden R, et al. A novel RT-PCR for the detection of *Helicobacter pylori* and identification of clarithromycin resistance mediated by mutations in the 23S rRNA gene. Diagnostic Microbiology and Infectious Disease. 2018;**90**(1):1-6

[26] Dadashzadeh K, Milani M, Rahmati M, et al. Real-time PCR detection of 16S rRNA novel mutations associated with *Helicobacter pylori* tetracycline resistance in Iran. Asian Pacific Journal of Cancer Prevention. 2014;**15**(20):8883-8886

[27] Pastukh N, Binyamin D, On A, et al. GenoType® HelicoDR test in comparison with histology and culture for *Helicobacter pylori* detection and identification of resistance mutations to clarithromycin and fluoroquinolones. Helicobacter. 2017;**22**(6):e12447

[28] Draper JL, Hansen LM, Bernick DL, et al. Fallacy of the unique genome: Sequence diversity within single *Helicobacter pylori* strains. Fraser CM, editor. MBio. 2017;**8**(1):e02321-e02316

[29] Megraud F, Coenen S, Versporten A, et al. *Helicobacter pylori* resistance to antibiotics in Europe and its relationship to antibiotic consumption. Gut. 2013;**62**(1):34-42

[30] Gao W, Cheng H, Hu F, et al. The evolution of *Helicobacter pylori* antibiotics resistance over 10 years in Beijing, China. Helicobacter. 2010;**15**:460-466

[31] Selgrad M, Meile J, Bornschein J, et al. Antibiotic susceptibility of *Helicobacter pylori* in central Germany and its relationship with the number of eradication therapies. European Journal of Gastroenterology & Hepatology. 2013;**25**(11):1257-1260

[32] Almeida N, Romãozinho JM, Donato MM, et al. *Helicobacter pylori* antimicrobial resistance rates in the central region of Portugal. Clinical Microbiology and Infection. 2014;**20**(11): 1127-1133

[33] Ogata SK, Gales AC, Kawakami E. Antimicrobial susceptibility testing for *Helicobacter pylori* isolates from Brazilian children and adolescents: Comparing agar dilution, e-test, and disk diffusion. Brazilian Journal of Microbiology. 2014;**45**(4):1439-1448

[34] Ahmad N, Zakaria WR, Mohamed R. Analysis of antibiotic susceptibility patterns of *Helicobacter pylori* isolates from Malaysia. Helicobacter. 2011;**16**:47-51

[35] Ang TL, Fock KM, Ang D, et al. The changing profile of *Helicobacter pylori* antibiotic resistance in Singapore: A 15-year study. Helicobacter. 2016;**21**(4):261-265

[36] Binh TT, Shiota S, Nguyen LT, et al. The incidence of primary antibiotic resistance of *Helicobacter pylori* in Vietnam. Nixon AE, editor. Journal of Clinical Gastroenterology. 2013; **47**(3):233-238

[37] Boehnke KF, Valdivieso M, Bussalleu A, et al. Antibiotic resistance among *Helicobacter pylori* clinical isolates in Lima, Peru. Infection and Drug Resistance. 2017;**10**:85-90

[38] Bouihat N, Burucoa C, Benkirane A, et al. *Helicobacter pylori* primary antibiotic resistance in 2015 in Morocco: A phenotypic and genotypic prospective and multicenter study. Microbial Drug Resistance. 2016;**23**(6):727-732

[39] Caliskan R, Tokman HB, Erzin Y, et al. Antimicrobial resistance of *Helicobacter pylori* strains to five antibiotics, including levofloxacin, in Northwestern Turkey. Revista da Sociedade Brasileira de Medicina Tropical. 2015;**48**(3):278-284

[40] Chang WL, Sheu BS, Cheng HC, et al. Resistance to metronidazole, clarithromycin and levofloxacin of *Helicobacter pylori* before and after clarithromycin-based therapy in Taiwan. Journal of Gastroenterology and Hepatology. 2009;**24**(7):1230-1235

[41] Cheng A, Sheng WH, Liou JM, et al. Comparative in vitro antimicrobial susceptibility and synergistic activity of antimicrobial combinations against *Helicobacter pylori* isolates in Taiwan. Journal of Microbiology, Immunology, and Infection. 2015;**48**(1):72-79

[42] Cuadrado-Lavín A, Salcines-Caviedes JR, Carrascosa MF, et al. Antimicrobial susceptibility of *Helicobacter pylori* to six antibiotics currently used in Spain. The Journal of Antimicrobial Chemotherapy. 2012;**67**(1):170-173

[43] Ducournau A, Bénéjat L, Sifré E, et al. *Helicobacter pylori* resistance to antibiotics in 2014 in France detected by phenotypic and genotypic methods. Clinical Microbiology and Infection. 2016;**22**(8):715-718

[44] Eisig JN, Silva F, Barbuti RC, et al. *Helicobacter pylori* antibiotic resistance in Brazil: Clarithromycin is still a good option. Arquivos de Gastroenterologia. 2011;**48**(4):261-264

[45] Farshad S, Alborzi A, Japoni A, et al. Antimicrobial susceptibility of *Helicobacter pylori* strains isolated from patients in Shiraz, Southern Iran. World Journal of Gastroenterology. 2010;**16**(45):5746-5751

[46] Dargiene G, Kupcinskas J, Jonaitis L, et al. Primary antibiotic resistance of *Helicobacter pylori* strains among adults and children in a tertiary referral centre in Lithuania. APMIS. 2017

[47] Goh KL, Navaratnam P. High *Helicobacter pylori* resistance to metronidazole but zero or low resistance to clarithromycin, levofloxacin, and other antibiotics in Malaysia. Helicobacter. 2011;**16**(3):241-245

[48] Gościniak G, Biernat M, Grabińska J, et al. The antimicrobial susceptibility of *Helicobacter pylori* strains isolated from children and adults with primary infection in the Lower Silesia Region, Poland. Polish Journal of Microbiology. 2014;**63**(1):57-61

[49] Gunnarsdottir AI, Gudjonsson H, Hardardottir H, et al. Antibiotic susceptibility of *Helicobacter pylori* in Iceland. Infectious Diseases. (Auckland). 2017;**49**(9):647-654

[50] Karczewska E, Wojtas-Bonior I, Sito E, et al. A primary and secondary clarithromycin, metronidazole, amoxicillin and levofloxacin resistance to *Helicobacter pylori* in southern Poland. Pharmacological Reports. 2011;**63**(3):799-807

[51] Kostamo P, Veijola L, Oksanen A, et al. Recent trends in primary antimicrobial resistance of *Helicobacter pylori* in Finland. International Journal of Antimicrobial Agents. 2011;**37**(1): 22-25

[52] Kupcinskas L, Rasmussen L, Jonaitis L, et al. Evolution of *Helicobacter pylori* susceptibility to antibiotics during a 10-year period in Lithuania. APMIS. 2013;**121**(5):431-436

[53] Larsen AL, Ragnhildstveit E, Moayeri B, et al. Resistance rates of metronidazole and other antibacterials in *Helicobacter pylori* from previously untreated patients in Norway. APMIS. 2013;**121**(4):353-358

[54] Ben Mansour K, Burucoa C, Zribi M, et al. Primary resistance to clarithromycin, metronidazole and amoxicillin of *Helicobacter pylori* isolated from Tunisian patients with peptic ulcers and gastritis: a prospective multicentre study. Annals of Clinical Microbiology and Antimicrobials. 2010;**9**(1):22

[55] Miftahussurur M, Syam AF, Nusi IA, et al. Surveillance of *Helicobacter pylori* antibiotic susceptibility in Indonesia: Different resistance types among regions and with novel genetic mutations. PLoS One. 2016;**11**(12):1-17

[56] O'Connor A, Taneike I, Nami A, et al. *Helicobacter pylori* resistance rates for levofloxacin, tetracycline and rifabutin among Irish isolates at a reference centre. Irish Journal of Medical Science. 2013:1-3

[57] Quek C, Pham ST, Tran KT, et al. Antimicrobial susceptibility and clarithromycin resistance patterns of *Helicobacter pylori* clinical isolates in Vietnam. F1000Research. 2016;**5**(0):671

[58] Raymond J, Lamarque D, Kalach N, et al. High level of antimicrobial resistance in French *Helicobacter pylori* isolates. Helicobacter. 2010;**15**(1):21-27

[59] Saracino IM, Zullo A, Holton J, et al. High prevalence of primary antibiotic resistance in *Helicobacter pylori* isolates in Italy. Journal of Gastrointestinal and Liver Diseases. 2012;**21**(4):363-365

[60] Seck A, Burucoa C, Dia D, et al. Primary antibiotic resistance and associated mechanisms in *Helicobacter pylori* isolates from Senegalese patients. Annals of Clinical Microbiology and Antimicrobials. 2013;**12**:3

[61] Shiota S, Reddy R, Alsarraj A, et al. Antibiotic resistance of *Helicobacter pylori* among male United States veterans. Clinical Gastroenterology and Hepatology. 2015;**13**(9):1616-1624

[62] Teh X, Khosravi Y, Lee WC, et al. Functional and molecular surveillance of *Helicobacter pylori* antibiotic resistance in Kuala Lumpur. PLoS One. 2014;**9**(7)

[63] Torres-Debat ME, Pérez-Pérez G, Olivares A, et al. Antimicrobial susceptibility of *Helicobacter pylori* and mechanisms of clarithromycin resistance in strains isolated from patients in Uruguay. Revista Española de Enfermedades Digestivas. 2009;**101**(11):757-762

[64] Korn VR, Gumnarai P, Ratanachu-ek T, et al. Nationwide survey of *Helicobacter pylori* antibiotic resistance in Thailand. Diagnostic Microbiology and Infectious Disease. 2013; **77**(4):346-349

[65] Wuppenhorst N, Draeger S, Stuger HP, et al. Prospective multicentre study on antimicrobial resistance of *Helicobacter pylori* in Germany. The Journal of Antimicrobial Chemotherapy. 2014;**69**(11):3127-3133

[66] Zollner-Schwetz I, Leitner E, Plieschnegger W, et al. Primary resistance of *Helicobacter pylori* is still low in Southern Austria. International Journal of Medical Microbiology. 2016; **306**(4):206-211

[67] Wu IT, Chuah SK, Lee CH, et al. Five-year sequential changes in secondary antibiotic resistance of *Helicobacter pylori* in Taiwan. World Journal of Gastroenterology. 2015;**21**(37): 10669-10674

[68] Oleastro M, Cabral J, Ramalho PM, et al. Primary antibiotic resistance of *Helicobacter pylori* strains isolated from Portuguese children: A prospective multicentre study over a 10 year period. The Journal of Antimicrobial Chemotherapy. 2011;**66**(10):2308-2311

[69] Van Boeckel TP, Gandra S, Ashok A, et al. Global antibiotic consumption 2000 to 2010: An analysis of national pharmaceutical sales data. The Lancet Infectious Diseases. 2014; **14**(8):742-750

[70] WHO. Antimicrobial Resistance. Global Report on Surveillance. Geneva: World Health Organization; 2014. pp. 383-394

[71] Hartzen SH, Andersen LP, Bremmelgaard A, et al. Antimicrobial susceptibility testing of 230 *Helicobacter pylori* strains: Importance of medium, inoculum, and incubation time. Antimicrobial Agents and Chemotherapy. 1997;**41**(12):2634-2349

[72] Rasmussen L. *Helicobacter pylori* [PhD thesis]. Copenhagen; 2013

[73] Petersen AM, Gjøde P, Vinge OD, et al. *Helicobacter pylori* antimicrobial resistance and risk factors in Denmark 1998-2004: No need for concern? Helicobacter. 2006;**11**(3):210-211

[74] Selgrad M, Tammer I, Langner C, et al. Different antibiotic susceptibility between antrum and corpus of the stomach, a possible reason for treatment failure of *Helicobacter pylori* infection. World Journal of Gastroenterology. 2014;**20**(43):16245-16251

[75] Choudhari SP, Pendleton KP, Ramsey JD, et al. A systematic approach toward stabilization of CagL, a protein antigen from *Helicobacter Pylori* that is a candidate subunit vaccine. Journal of Pharmaceutical Sciences. 2013;**102**:2508-2519

[76] Kwok T, Zabler D, Urman S, et al. *Helicobacter* exploits integrin for type IV secretion and kinase activation. Nature. 2007;**449**(7164):862-866

[77] Chionh YT, Arulmuruganar A, Venditti E, et al. Heat shock protein complex vaccination induces protection against *Helicobacter pylori* without exogenous adjuvant. Vaccine. 2014;**32**:2350-2358

[78] Nedrud JG, Bagheri N, Schön K, et al. Subcomponent vaccine based on CTA1-DD adjuvant with incorporated UreB class II peptides stimulates protective *Helicobacter pylori* immunity. Ho PL, editor. PLoS One. 2013;**8**(12):e83321

[79] D'Elios MM, Czinn SJ. Immunity, Inflammation, and Vaccines for *Helicobacter pylori*. Heli cobacter. 2014;**19**(S1):19-261

[80] Zeng M, Mao XH, Li JX, et al. Efficacy, safety, and immunogenicity of an oral recombinant *Helicobacter pylori* vaccine in children in China: A randomised, double-blind, placebo-controlled, phase 3 trial. Lancet. 2015;**386**(10002):1457-1464

[81] Chen M, Jensen B, Zhai L, et al. Nizatidine and omeprazole enhance the effect of metronidazole on *Helicobacter pylori* in vitro. International Journal of Antimicrobial Agents. 2002;**19**(3):195-200

[82] Kristiansen JE, Justesen T, Hvidberg EF, et al. Trimipramine and other antipsychotics inhibit *Campylobacter pylori* in vitro. Pharmacology & Toxicology. 1989;**64**(4):386-388

[83] Fahey JW, Stephenson KK, Wallace AJ. Dietary amelioration of *Helicobacter* infection. Nutrition Research. 2015;**35**(6):461-473

[84] Rahbar M, Mardanpur K, Tavafzadeh R. Imprint cytology: A simple, cost effectiveness analysis for diagnosing *Helicobacter pylori*, in west of Iran. Medical journal of the Islamic Republic of Iran. 2012;**26**(1):12-16

[85] Warburton-Timms V, McNulty C. Role of screening agar plates for in vitro susceptibility testing of *Helicobacter pylori* in a routine laboratory setting. Journal of Clinical Pathology. 2001;**54**(5):408-411

Chapter 8

Nonantibiotic-Based Therapeutics Targeting *Helicobacter pylori*: From Nature to the Lab

Paula Parreira, Catarina Leal Seabra,
Daniela Lopes-de-Campos and
Maria Cristina L. Martins

Additional information is available at the end of the chapter

http://dx.doi.org/10.5772/intechopen.81248

Abstract

The available therapy against *Helicobacter pylori* is based on a combination of antibiotics and proton pump inhibitors. The high prevalence of antibiotic-resistant strains leads to failure of this complex therapeutic regimen, leaving millions of people worldwide without effective therapeutic options. "Nature-derived" bioactive compounds with antibacterial performance may be of value for developing newer and more effective strategies. For centuries, natural compounds have played a pivotal role in traditional medicine and, in the last decades, they have gained renewed strength in the clinical field, boosted by advances in chemical characterization and extensive activity screening. Also, their recognition in gastric infection management has been empowered by the bioengineering field, namely by the development of stomach-specific delivery strategies. In this chapter, natural bioactive compounds, such as polyunsaturated fatty acids and triterpenic acids with anti-*H. pylori* effect, are described. The bioengineering approaches used to overcome their limited intrinsic bioavailability are briefly highlighted.

Keywords: nanotechnology, bioactive compounds, lipophilic compounds, phytochemicals, antibiotic-free therapies

1. Introduction

Helicobacter pylori is the etiologic agent of several gastric disorders that may range from chronic gastritis to more severe outcomes [1]. Ultimately, the complex interplay between *H. pylori*, the host susceptibility, and environmental factors such as smoking and drinking can lead to gastric cancer, which is the fifth most common cancer worldwide, accounting for 754,000 deaths in 2015 [2, 3]. Despite significant medical advances, the 5-year survival rate from gastric cancer is low (31%), mainly because this cancer is diagnosed at later stages [4]. It is widely recognized that the best strategy to reduce the risk of gastric carcinoma associated with *H. pylori* infection is its eradication from infected hosts [5, 6]. The current treatment relies on a combination of antibiotics (clarithromycin plus amoxicillin or metronidazole) and an acid-suppressive drug (e.g., proton-pump inhibitor), since no available substances are effective as monotherapy [7]. However, the eradication rates of this therapeutic scheme have been declining to unacceptable levels [8], mostly due to high antibiotic resistance levels. In fact, *H. pylori* has been placed among the 16 antibiotic-resistant bacteria that pose greatest threat to human health [9]. It is noteworthy that besides resistance and coinfection with multiple strains with distinct antibiotic susceptibilities, other factors also account for conventional treatment failure:

a. drugs bioavailability: antibiotics are typically administrated via the oral route. Since the gastric mucus layer acts as a barrier to antibiotic delivery, most drugs are unable to reach the underlying gastric epithelium, where *H. pylori* is attached [10];

b. gastric features: the pH at the stomach/duodenum varies from acidic to neutral, depending on the presence or absence of food and the location within the mucus barrier that covers the epithelial cells. This affects the efficacy of most antibiotics once only a few remain active in a wide pH range [11–13];

c. compliance: side effects, such as taste disturbances with clarithromycin and metronidazole, and diarrhea with amoxicillin, account for the poor patient compliance. Additionally, complex regimens that require multiple doses of medication each day also decrease therapeutic compliance [14];

d. lifestyle: smoking and alcohol consumption are thought to contribute for treatment failure [15].

In this scenario, where *H. pylori* traditional antibiotic therapies fail, a considerable interest in alternative therapeutic players combined with bioengineering strategies has arisen. Several "nature-derived" options have been studied [5], and some of them are summarized in **Figure 1**.

Although very promising *in vitro*, these molecules share a common drawback when transferred to *in vivo* settings: intrinsic limited bioavailability. **Figure 2** summarizes some of the bioengineering approaches envisioned to overcome the mentioned limitation.

Chitosan micro-/nanoparticles have been extensively studied as drug delivery systems targeting *H. pylori* infection [16], mainly due to its gastric retentive properties [17]. Chitosan, a natural biocompatible polysaccharide obtained by *N*-deacetylation of chitin [18], has mucoadhesive properties due to electrostatic interactions between its cationic amine groups and gastric mucins, which are negatively charged at the acidic gastric pH [19, 20].

Chitosan microspheres were used to encapsulate and improve the biological activity of trans-resveratrol, a phenolic compound that has, among other biological activities, anti-*H. pylori*

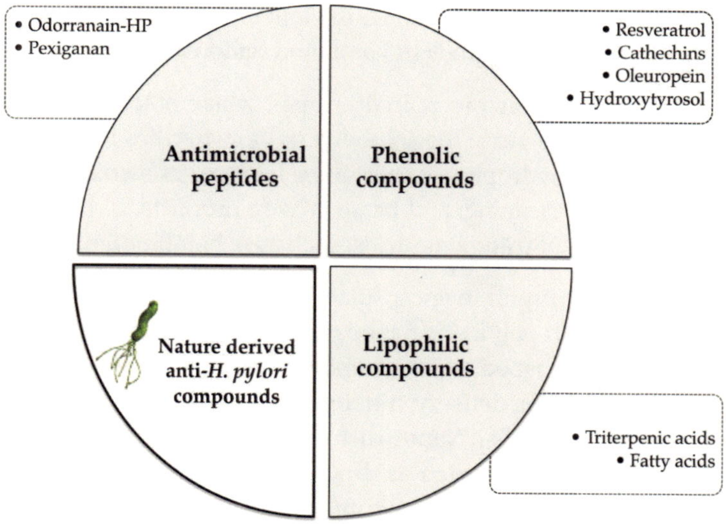

Figure 1. Some "nature-derived" bioactive compounds described in the literature in the scope of novel anti-*H. pylori* therapeutics.

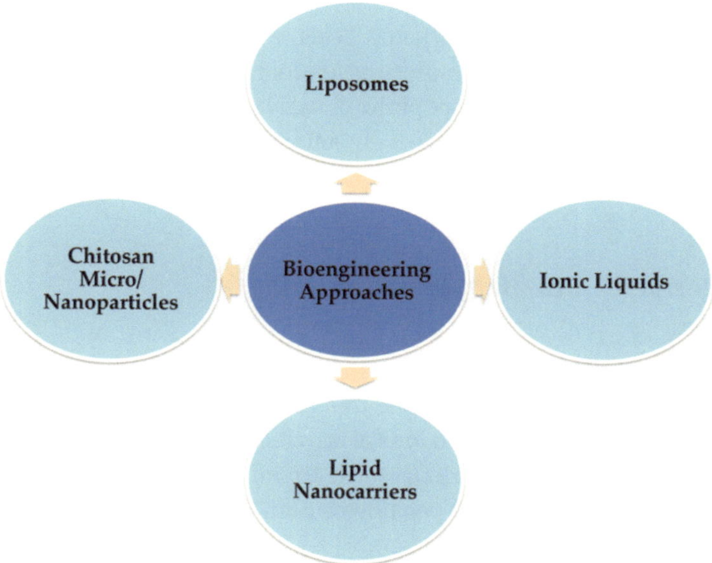

Figure 2. Most common bioengineering approaches applicable to "nature-derived" bioactives in the scope of *H. pylori* infection management.

action [21]. Polyelectrolyte complex nanoparticles (PNPs), prepared by the combination of chitosan with negatively charged polymers, alginate or heparin, have also been used to encapsulate the antimicrobial peptide pexiganan [22] and berberine [23], respectively. This approach increased the effectiveness of these bioactive compounds, inhibiting *H. pylori* growth and reducing their cytotoxic effects.

Other strategies that can be used to overcome the low bioavailability and solubility of lipophilic compounds comprise ionic liquids (IL) and lipid nanoparticles.

IL are a new class of powerful catanionic hydrotropes, where both the cation and the anion synergistically contribute to increase the solubility of biomolecules in water [24]. Therefore, IL enhance the solubility of hydrophobic substances in aqueous media and are widely used in the formulation of drugs, cleaning, and personal care products. IL have been explored to increase the water solubility of triterpenic acids, such as of betulinic acid [25, 26].

Lipid nanoparticles are very useful to encapsulate lipophilic compounds due to their higher biocompatibility compared to polymeric nanoparticles [27, 28]. Liposomes, firstly described in the mid-1960s, are sphere-shaped vesicles consisting of one or more phospholipid bilayers. Liposomes are the first nanodrug delivery systems that have been successfully translated into real-time clinical application [29–31]. Nanostructured lipid carriers (NLCs) are lipid nanoparticles specifically designed and patented as drug delivery systems, and they are characterized by a solid-lipid core composed of a mixture of solid and liquid lipids [32]. NLCs can be prepared using a wide variety of lipids including fatty acids, glyceride mixtures, or waxes, stabilized with biocompatible surfactants, which makes this a very versatile strategy. Both are considered safe and under the Generally Recognized as Safe (GRAS) status issued by the Food and Drug Administration (FDA) [31, 33].

Polyunsaturated fatty acids and triterpenic acids with anti-*H. pylori* effect have gained renewed interest in the scientific community as alternatives to overcome the increasing number of drug-resistant bacteria. These lipophilic bioactive compounds can largely benefit from a nanotechnological approach to improve their stability and to overcome their limited intrinsic bioavailability and thus, they will be briefly highlighted in the next sections.

2. "Nature-derived" anti-*H. pylori* fatty acids

Free fatty acids, also known as antimicrobial lipids [34], are linear carbon chains, which are the main constituent of phospholipids, triglycerides, sterol esters, among others [35]. Consequently, they are important for biological activities, such as for energetic, metabolic, and structural processes [35]. Fatty acids are classified according to the length of the carbon chains, the number of double bonds, and their positions within the moiety [35]. Polyunsaturated fatty acids (PUFAs) have two or more double bonds [36], and they have been recognized for their broad-spectrum activity against bacteria (e.g., *H. pylori*), fungi, protozoa, and virus [36–38]. This is due to the ability of fatty acids to work as mild surfactants [34]. The disturbance of the bacterial cell membrane can lead to the deregulation of metabolic pathways, inhibiting the bacterial growth, or even to lysis and death [34]. The specific interaction between antimicrobial fatty acids and bacterial membranes remains to be fully understood. Some studies revealed that free fatty acids can induce different kinds of morphological changes in the membrane [39, 40]. Khulusi et al. and Correia et al. reported that fatty acids can be incorporated into *H. pylori* phospholipids membrane, being able to change the bacillary morphology of the bacteria to their coccoid shape [32, 41–43]. Several studies also identified the bacterial membrane as the main target of fatty acids, leading to a sequence of biophysical phenomena including membrane destabilization, pore formation, and lysis of bacteria [32, 41, 44, 45]. The multiple mechanisms that are behind their ability to perturb bacterial cell membranes lead to a low probability of antimicrobial resistance [34]. On the opposite, small molecules, such as commercial antibiotics, inhibit specific enzymes and, consequently, increase the probability of antimicrobial resistance development [34].

The antibacterial activity of PUFAs depends on their molecular structure. In fact, the existence of double bonds and, more specifically, their number and orientation within the fatty acids are important for their physicochemical properties [46, 47]. These structural differences are reported to affect their ability to inhibit *H. pylori* growth *in vitro* [45]. For instance, the inhibitory effect is higher for higher degrees of unsaturation: [oleic (C18:1) < linoleic (C18:2) < arachidonic (C20:4) < n-3 linolenic (C18:3) = n-6 linolenic (C18:3) = eicosapentaenoic (C20:5) acid] [45]. The encapsulation of free fatty acids in nanoparticles can improve their pharmacokinetic and pharmacological properties [34]. A review of the nanotechnology formulations used to encapsulate antimicrobial lipids was recently published [34].

In this section, two of the most promising fatty acids (docosahexaenoic acid (DHA) and linolenic acid (LA)) are described regarding their specific application against *H. pylori*.

Figure 3 illustrates the two bioengineering approaches that have been applied to DHA and LA that will be discussed in the following subsections.

2.1. Docosahexaenoic acid (DHA)

DHA inhibits *H. pylori* growth both *in vitro* and *in vivo*, since it is able to reduce *H pylori* adhesion to gastric cells and bacterial ATP production [42, 43, 48]. DHA induces changes in expression of *H. pylori* outer membrane proteins associated with stress response, metabolism, and modified bacterial lipopolysaccharide phenotype [43]. DHA is also able to indirectly interfere with *H. pylori* growth since it alters cholesterol levels in epithelial cells, thereby influencing the bacterium ability to uptake and use epithelial cholesterol [48].

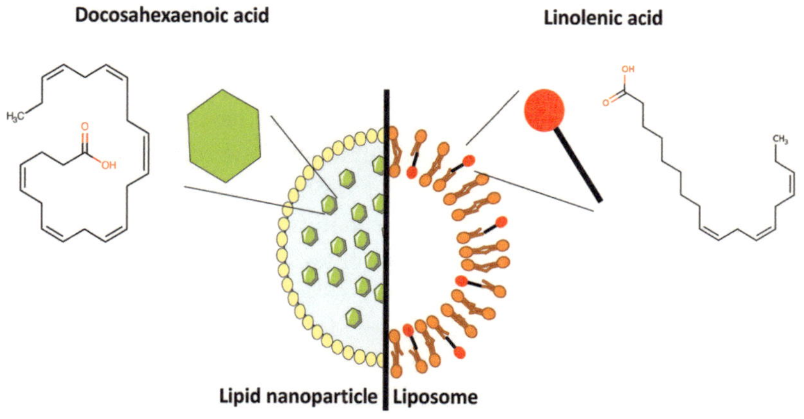

Figure 3. Encapsulation of fatty acids, namely DHA and LA, in different types of nanoparticles.

Although *in vivo* studies using gastric-infected mice demonstrated that DHA was able to decrease only 50% of *H. pylori* gastric colonization, the DHA conjugation with antibiotic standard treatment decreased the recurrence of *H. pylori* infection [42, 43].

Another DHA feature is its ability to attenuate the host inflammatory response associated with gastric infection [49, 50].

DHA poor solubility in water, fast oxidation/degradation plus gastric settings drawbacks (namely low gastric residence time and low penetration through the gastric mucus layer) are challenging issues for its clinical translation [36, 50, 51]. To overcome these obstacles, cytocompatible lipid nanoparticles have been researched to encapsulate DHA [44, 52]. It was demonstrated that DHA lipid nanoparticles are able to destabilize *H. pylori* membranes, leading to disruption and leakage of cytoplasmic contents [32, 44]. Importantly, these lipid nanoparticles do not interfere with normal gut microbiota in opposite to dramatic changes described for the conventional antibiotic therapy [44].

2.2. Linolenic acid (LA)

LA, as fatty acids in general, is considered safe [53]. It is classified as an essential fatty acid, once it cannot be synthesized by the human body, being necessary to be supplied by the diet [54]. Its importance for biological processes is unquestionable. LA undergoes metabolic changes *in vivo* that ultimately lead to the formation of prostaglandins, thromboxanes, leukotrienes, and lipoxins [54]. Furthermore, the usefulness of LA as an antibacterial agent was also proved, being one of the most potent unsaturated fatty acids against *H. pylori* [7]. It also promotes the adhesion of *Lactobacillus casei* to mucosa surfaces, which indirectly hinders the growth of *H. pylori* [55]. Besides its bactericidal effect, LA is also important for the integrity of the gastric mucosa. It was already proposed that lower levels of essential fatty acids, such as LA, lead to decreased levels of prostaglandins and, consequently, to a higher susceptibility of the gastric mucosa to ulcerogenic agents [56].

Nanotechnology has been successfully used to load fatty acids, including LA [34]. As above mentioned, the oral administration of fatty acids is hindered by their poor solubility, especially at acidic pH, and their susceptibility to chemical degradation [57]. In fact, the carboxyl protonation under acidic pH at the stomach lumen decreases the efficacy of fatty acids after oral administration [53]. This was already shown *in vivo*, with no significant effect of plain LA in killing *H. pylori* on a mouse model [53]. Nevertheless, liposomes are promising bioengineering strategies to overcome these limitations. Due to the amphiphilic nature of fatty acids, they can be easily incorporated into the phospholipid bilayer of liposomes [57]. Hence, Obonyo et al. used liposomes of egg phosphatidylcholines, cholesterol, and LA to kill *H. pylori* [57]. They showed that LA-loaded liposomes were effective against *H. pylori* even in its coccoid form and regardless their resistance to antibiotics [57]. Interestingly, *H. pylori* developed resistance against free LA at subbactericidal concentrations, whereas it showed no resistance against LA when incorporated into the nanoparticles [57]. These results show the promising usefulness of nanotechnology not only to protect the fatty acid from its degradation, but also to improve its efficacy. The higher efficacy relies on their ability to fuse with the

http://dx.doi.org/10.5772/intechopen.81248

bacterial membrane, being directly and faster incorporated into the bacterial membrane [57]. Their main mechanism is the increase of the permeability of the outer membrane and of the plasma membrane of *H. pylori,* which leads to a leakage of cytoplasmic contents [58]. They also decrease the *H. pylori*-induced proinflammatory cytokines, helping in the healing of the gastric mucosa [58]. Furthermore, the ability of those liposomes to be retained at the site of infection was also shown *in vivo* [53]. The retention for up to 24 h was attributed to the small size of the liposomes and their anionic surface charge, which decreases their hydrophobic entrapment [53]. The biocompatibility of the formulation was shown through gastric histopathology and mucosal integrity and by the maintenance of the gut microbiota, on the opposite of the current triple therapy [53, 59].

3. "Nature-derived" anti-*H. pylori* phytochemicals

Phytochemicals have been used for centuries in the treatment of gastrointestinal disorders, such as dyspepsia, gastritis, and peptic ulcer disease [60]. Over the last two decades, phytotherapy has gained strength in the scientific community, prompted by the need of alternatives to the ineffectiveness of traditional antibiotics.

Plants synthesize a vast range of secondary metabolites with a significant portion consisting of phenolic and flavonoid compounds [61]. These secondary metabolites, other than providing plants with unique survival or adaptive strategies, are associated to a wide range of biological activities [62]. Phenolic compounds, namely wine polyphenols, from which resveratrol is the most studied, and olive oil polyphenols, mainly hydroxytyrosol, have been associated with anti-*H. pylori* activity [5]. Lipophilic compounds from the terpenes family can also be obtained from several plants. In the scope of anti-*H. pylori* strategies, these are described in more detail in the following section.

3.1. Triterpenic acids

Terpenes are naturally occurring hydrocarbons, with the general formula $(C_5H_8)_n$ ($-(-CH_2=C(CH_3)-CH=CH_2)_n$), where n is the number of isoprene units. Depending on the number of isoprene building blocks, they are classified into several groups, such as monoterpenes, sesquiterpenes, diterpenes, triterpenes, and tetraterpenes (with 2, 3, 4, 6, and 8 isoprene units, respectively). These compounds can undergo chemical modifications by oxidation or rearrangement of the carbon skeleton, which leads to a vast group of compounds denominated terpenoids [63].

Pentacyclic triterpenoids are commonly isolated as active substances from different natural sources, mainly plant surfaces such as stem bark or leaf and fruit waxes [64]. Among them, pentacyclic triterpenes ($C_{30}H_{48}$) are being marketed as therapeutic agents or dietary supplements around the world due to their biological applications [65, 66]. Their antibacterial properties are also recognized. For instance, it was demonstrated that the acidic fraction of the total mastic extract without polymer (TMEWP) from the Chios Mastic Gum (resin of

Pistacia lentiscus var. chia) is effective in killing *H. pylori* [67]. This antibacterial effect was attributed to their rich composition in oleanolic acid, isomasticadienolic acid, masticadienolic, and moronic acid [67]. Paraschos et al. demonstrated that the prophylactic treatment with the TMEWP was not able to prevent *H. pylori* infection in C57BL/6 mice infected with mouse-adapted *H. pylori* SS1 strain [67]. Nevertheless, the number of *H. pylori* colonies significantly reduced (1.5 log colony forming units/g of tissue) when the animals were subjected to continuous administration of 0.75 mg of TMEWP for 3 months [67]. Shin et al. reported that betulinic acid and oleanolic acid, extracted from *Fosythia suspensa,* were able to inhibit the urease activity of *H. pylori* ATCC 43504 [68]. Furthermore, Parreira et al. reported that outer bark extracts of *Eucalyptus nitens* and *E. globulus,* rich in betulinic, betulonic, oleanolic, and ursolic acids (**Figure 4**), have anti-*H. pylori* activity against strains with distinct virulence degree [69]. Interestingly, the eucalyptus extracts had a lower minimal inhibitory concentration than the isolated pure triterpenic acids, which led to the conclusion that the final observed antibacterial effect was due to synergic effects [69].

Although not specifically designed toward *H. pylori* infection, different strategies to improve the oral bioavailability of triterpenic acids have been studied. For example, oleanolic acid bioavailability has been enhanced by using a phospholipids complex with hydroxyapatite [70]. Yang et al. have developed liposomes to increase ursolic acid bioavaliabitliy [71] and pharmacokinetic studies carried out by Ge et al. reported that the oral bioavailability of ursolic acid was 27.5-fold higher when it was incorporated in nanoparticles than when administered as a free compound [72].

The abovementioned advances in increasing the bioavailability of triterpenic acids using bioengineering strategies will enable, in the near future, to further pursue research of novel nonantibiotic and more effective "nature-inspired" therapies against *H. pylori.*

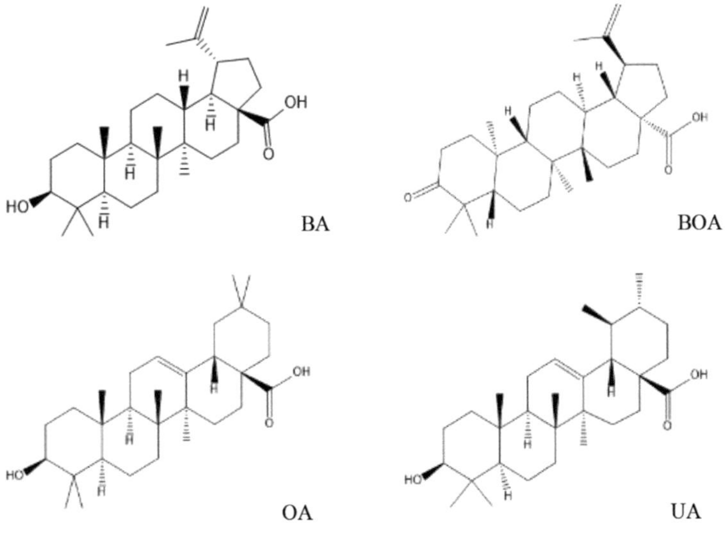

Figure 4. Chemical structures of triterpenic acids: betulinic (BA), betulonic (BOA), oleanolic (OA), and ursolic (UA) acids.

4. Translation to real-world scenario

Both fatty acids and triterpenic acids have been reported to exhibit similar performance against *H. pylori*. Nevertheless, their action mechanisms are fairly distinct: while fatty acids are reported to interact with the bacterial membrane, triterpenic acids are reported to be more involved in enzymatic inhibition, namely urease hindering [5]. Since both bioactives classes target crucial structures for *H. pylori* survival, emergence of resistance is not anticipated, as it would require massive bacterial energy [32, 69].

Despite the remarkable effects associated to fatty and triterpenic acids for gastric infection management, translation into real-world applications is still delayed. For that, it has contributed the fact that only in the last decade more attention has been paid to nature-derived molecules, counteracting the "chemical pharmacological" tendency that had been initiated in the beginning of the twentieth century. Also, there was a significant reduction of investment in the clinical development of antibiotics over the last years. In fact, only 1.6% of the drugs under clinical development by the world's largest drug companies are antibiotics [73]. This has boosted the search for other sources of antimicrobials. In addition, bioengineering emerged in the twenty-first century as a powerful tool to develop drug delivery systems and, consequently, to overcome the more generalized drawbacks associated with the lipophilic bioactive compounds discussed in this chapter [5]. Bioengineering approaches for fatty acids specific application against *H. pylori* are already on a "fast-track," while those for triterpenic acids are only now evolving, which explains the lack of solid studies coupling these bioactives with bioengineering strategies.

To the date and to the best of our knowledge, most of the herein described compounds are in *in vivo* studies phase, being expected that in the next few years some will cross the clinical trials barrier. There are several factors contributing to the anticipated success of these "nature-based" strategies. They are generally cost-effective, due to their abundance in nature, and they require low-cost extraction productions. Furthermore, the biotechnological improvements that include nanotechnological coupling to nature-derived molecules will hopefully contribute to reaching "real-life" applications. In addition, more "nature-based" molecules are reaching the market with FDA approval to treat infectious disease, such as antimalaria Artemisinin therapies, based on an herb employed in Chinese traditional medicine [74], which anticipates the future success of nature-inspired strategies for *H. pylori* eradication.

5. Conclusion

H. pylori infection is one of the most prevalent infections worldwide, which is also reflected onto the high prevalence of gastric cancer. Emerging antibiotic resistance leads to an urgent need of alternative treatments. Resourcing to widely available lipophilic natural bioactive compounds with anti-*H. pylori* activity, namely fatty or triterpenic acids, should be further considered as novel therapeutic options. In this context, nanotechnology emerges as a key player, as it allows overcoming the bioactive major drawbacks that have been holding back their "real-world" application.

Acknowledgements

The authors acknowledge FEDER—Fundo Europeu de Desenvolvimento Regional funds through the COMPETE 2020—Operational Program for Competitiveness and Internationalization (POCI), Portugal 2020, NORTE-01-0145-FEDER-000012, and PyloriBinders—*Helicobacter pylori*-specific biomaterials for antibiotic-free treatment/diagnostic of gastric infection (PTDC/CTM-BIO/4043/2014).

Conflict of interest

The authors declare no conflict of interest.

Author details

Paula Parreira[1,2], Catarina Leal Seabra[1,2], Daniela Lopes-de-Campos[3] and Maria Cristina L. Martins[1,2,4]*

*Address all correspondence to: cmartins@ineb.up.pt

1 i3S—Instituto de Investigação e Inovação em Saúde, University of Porto, Porto, Portugal

2 INEB—Instituto de Engenharia Biomédica, University of Porto, Porto, Portugal

3 LAQV, REQUIMTE, Faculty of Pharmacy, University of Porto, Porto, Portugal

4 ICBAS—Instituto de Ciências Biomédicas Abel Salazar, University of Porto, Portugal

References

[1] Correa P, Houghton J. Carcinogenesis of *Helicobacter pylori*. Gastroenterology. 2007; **133**(2):659-672

[2] Ferlay J, Soerjomataram I, Ervik M, Dikshit R, Eser S, Mathers C, et al. GLOBOCAN 2012 v1.0, Cancer Incidence and Mortality Worldwide: IARC CancerBase. No. 11 [Internet]. Vol. 11. Lyon, France: International Agency for Research on Cancer; 2013. http://globocan.iarc.f

[3] WHO. WHO Cancer. WHO. 2017. Available from: http://www.who.int/mediacentre/factsheets/fs297/en/

[4] American Cancer Society. Cancer facts & figures 2017. American Cancer Society Journal. 2017;**2017**:1-71

[5] Parreira P, Fátima Duarte M, Reis CA, Martins MCL. *Helicobacter pylori* infection: A brief overview on alternative natural treatments to conventional therapy. Critical Reviews in Microbiology. 2016;**42**(1):94-105

[6] Piazuelo MB, Epplein M, Correa P. Gastric cancer: An infectious disease. Infectious Disease Clinics of North America. 2010;**24**(4):853-869

[7] Patel A, Shah N, Prajapati JB. Clinical appliance of probiotics in the treatment of *Helicobacter pylori* infection—A brief review. Journal of Microbiology, Immunology, and Infection. 2014;**47**(5):429-437

[8] Malfertheiner P, Megraud F, O'Morain CA, Gisbert JP, Kuipers EJ, Axon AT, et al. Management of *Helicobacter pylori* infection—The Maastricht V/Florence consensus report. Gut. 2017;**66**(1):6-30

[9] Dang BN, Graham DY. *Helicobacter pylori* infection and antibiotic resistance: A WHO high priority? Nature Reviews. Gastroenterology & Hepatology. 2017;**14**(7):383-384

[10] Ricci V, Zarrilli R, Romano M. Voyage of *Helicobacter pylori* in human stomach: Odyssey of a bacterium. Digestive and Liver Disease. 2002;**34**:2-8

[11] Erah PO, Goddard AF, Barrett DA, Shaw PN, Spiller RC. The stability of amoxycillin, clarithromycin and metronidazole in gastric juice: Relevance to the treatment of *Helicobacter pylori* infection. The Journal of Antimicrobial Chemotherapy. 1997;**39**:5-12

[12] Sherwood PV, Wibawa JID, Atherton JC, Jordan N, Jenkins D, Barrett DA, et al. Impact of acid secretion, gastritis, and mucus thickness on gastric transfer of antibiotics in rats. Gut. 2002;**51**(4):490-495

[13] Vakil N. *Helicobacter pylori* treatment: A practical approach. The American Journal of Gastroenterology. 2006;**101**(3):497-499

[14] Campo SMA, Zullo A, Hassan C, Morini S. Antibiotic treatment strategies for *Helicobacter pylori* infection. Recent Patents on Anti-Infective Drug Discovery. 2007;**2**(1):11-17

[15] Suzuki T, Matsuo K, Ito H, Sawaki A, Hirose K, Wakai K, et al. Smoking increases the treatment failure for *Helicobacter pylori* eradication. The American Journal of Medicine. 2006;**119**:217-224

[16] Gonçalves IC, Henriques PC, Seabra CL, Martins MCL. The potential utility of chitosan micro/nanoparticles in the treatment of gastric infection. Expert Review of Anti-Infective Therapy. 2014;**12**(8):981-992

[17] Fernandes M, Gonçalves IC, Nardecchia S, Amaral IF, Barbosa MA, Martins MCL. Modulation of stability and mucoadhesive properties of chitosan microspheres for therapeutic gastric application. International Journal of Pharmaceutics. 2013;**454**(1):116-124

[18] Sogias I, Williams AC, Khutoryanskiy VV. Why is chitosan mucoadhesive? Biomacromolecules. 2008;**9**:1837-1842

[19] Deacon MP, McGurk S, Roberts CJ, Williams PM, Tendler SJ, Davies MC, et al. Atomic force microscopy of gastric mucin and chitosan mucoadhesive systems. The Biochemical Journal. 2000;**348**(3):557-563

[20] Nogueira F, Gonçalves IC, Martins MCL. Effect of gastric environment on *Helicobacter pylori* adhesion to a mucoadhesive polymer. Acta Biomaterialia. 2012;**9**(2):5208-5215

[21] Altiok D, Altiok E, Bayraktar O, Tihminlioglu F. Stability of trans-resveratrol incorporated in chitosan microspheres. In: 14th Natl Biomed Eng Meet. 2009

[22] Zhang XL, Jiang AM, Ma ZY, Li XB, Xiong YY, Dou JF, et al. The synthetic antimicrobial peptide pexiganan and its nanoparticles (PNPs) exhibit the anti-*Helicobacter pylori* activity *in vitro* and *in vivo*. Molecules. 2015;**20**(3):3972-3985

[23] Chang C-H, Huang W-Y, Lai C-H, Hsu Y-M, Yao Y-H, Chen T-Y, et al. Development of novel nanoparticles shelled with heparin for berberine delivery to treat *Helicobacter pylori*. Acta Biomaterialia. 2011;**7**(2):593-603

[24] Cláudio AFM, Neves MC, Shimizu K, Canongia Lopes JN, Freire MG, Coutinho JAP. The magic of aqueous solutions of ionic liquids: Ionic liquids as a powerful class of catanionic hydrotropes. Green Chemistry. 2015;**17**(7):3948-3963

[25] Domínguez de María P, Maugeri Z. Ionic liquids in biotransformations: From proof-of-concept to emerging deep-eutectic-solvents. Current Opinion in Chemical Biology. 2011;**15**(2):220-225

[26] Ressmann AK, Strassl K, Gaertner P, Zhao B, Greiner L, Bica K. New aspects for biomass processing with ionic liquids: Towards the isolation of pharmaceutically active betulin. Green Chemistry. 2012;**14**(4):940

[27] Battaglia L, Gallarate M. Lipid nanoparticles: State of the art, new preparation methods and challenges in drug delivery. Expert Opinion on Drug Delivery. 2012;**9**(5):497-508

[28] Carbone C, Leonardi A, Cupri S, Puglisi G, Pignatello R. Pharmaceutical and biomedical applications of lipid-based nanocarriers. Journal of Pharmaceutical and Biomedical Analysis. 2014;**3**(2):199-215

[29] Takahashi M, Kitamoto D, Imura T, Oku H, Takara K, Wada K. Characterization and bioavailability of liposomes containing a ukon extract. Bioscience, Biotechnology, and Biochemistry. 2008;**72**(5):1199-1205

[30] Thamphiwatana S, Fu V, Zhu J, Lu D, Gao W, Zhang L. Nanoparticle-stabilized liposomes for ph-responsive gastric drug delivery. Langmuir. 2013;**29**(39):12228-12233

[31] Bulbake U, Doppalapudi S, Kommineni N, Khan W. Liposomal formulations in clinical use: An updated review. Pharmaceutics. 2017;**27**(9):2

[32] Seabra CL, Nunes C, Gomez-Lazaro M, Correia M, Machado JC, Gonçalves IC, et al. Docosahexaenoic acid loaded lipid nanoparticles with bactericidal activity against *Helicobacter pylori*. International Journal of Pharmaceutics. 2017;**519**(1-2):128-137

[33] Dolatabadi JEN, Valizadeh H, Hamishehkar H. Solid lipid nanoparticles as efficient drug and gene delivery systems: Recent breakthroughs. Advanced Pharmaceutical Bulletin. 2015;**5**(2):151-159

[34] Jackman JA, Yoon BK, Li D, Cho NJ. Nanotechnology formulations for antibacterial free fatty acids and monoglycerides. Molecules. 2016;**21**(3):305

[35] Chow CK. Fatty Acids in Foods and their Health Implications. 3rd ed. Boca Raton, FL: CRC Press: Taylor & Francis Group; 2008

[36] Desbois AP, Smith VJ. Antibacterial free fatty acids: Activities, mechanisms of action and biotechnological potential. Applied Microbiology and Biotechnology. 2010;**85**(6): 1629-1642

[37] Desbois AP. Potential applications of antimicrobial fatty acids in medicine, agriculture and other industries. Recent Patents on Anti-Infective Drug Discovery. 2012;**7**(2):111-122

[38] Yoon B, Jackman J, Valle-González E, Cho N-J. Antibacterial free fatty acids and monoglycerides: Biological activities, experimental testing, and therapeutic applications. International Journal of Molecular Sciences. 2018;**19**(4):1114

[39] Calder PC. Mechanisms of action of (n-3) fatty acids. The Journal of Nutrition. 2012;**142**(3): 592S-599S

[40] Sun CQ, O'Connor CJ, Roberton AM. Antibacterial actions of fatty acids and monoglycerides against *Helicobacter pylori*. FEMS Immunology and Medical Microbiology. 2003;**36**:9-17

[41] Khulusi S, Ahmed HA, Patel P, Mendall MA, Northfield TC. The effects of unsaturated fatty acids on *Helicobacter pylori in vitro*. Journal of Medical Microbiology. 1995;**42**(4):276-282

[42] Correia M, Michel V, Matos AA, Carvalho P, Oliveira MJ, Ferreira RM, et al. Docosahexaenoic acid inhibits helicobacter pylori growth in vitro and mice gastric mucosa colonization. PLoS One. 2012;**7**(4):e35072

[43] Correia M, Michel V, Osorio H, El Ghachi M, Bonis M, Boneca IG, et al. Crosstalk between *Helicobacter pylori* and gastric epithelial cells is impaired by docosahexaenoic acid. PLoS One. 2013;**8**(4):e60657

[44] Seabra CL, Nunes C, Brás M, Gomez-Lazaro M, Reis CA, Gonçalves IC, et al. Lipid nanoparticles to counteract gastric infection without affecting gut microbiota. European Journal of Pharmaceutics and Biopharmaceutics. 2018;**127**:378-386

[45] Thompson L, Cockayne A, Spiller RC. Inhibitory effect of polyunsaturated fatty acids on the growth of *Helicobacter pylori*: A possible explanation of the effect of diet on peptic ulceration. Gut. 1994;**35**:1557-1561

[46] Bazinet RP, Laye S. Polyunsaturated fatty acids and their metabolites in brain function and disease. Nature Reviews. Neuroscience. 2014;**15**(12):771-785

[47] Catal A. Five decades with polyunsaturated fatty acids: Chemical synthesis, enzymatic formation, lipid peroxidation and its biological effects. Journal of Lipids. 2013;**2013**:19

[48] Correia M, Casal S, Vinagre J, Seruca R, Figueiredo C, Touati E, et al. *Helicobacter pylori*'s cholesterol uptake impacts resistance to docosahexaenoic acid. International Journal of Medical Microbiology. 2014;**304**(3-4):314-320

[49] Park S-H, Kangwan N, Park J-M, Kim E-H, Hahm KB. Non-microbial approach for *Helicobacter pylori* as faster track to prevent gastric cancer than simple eradication. World Journal of Gastroenterology. 2013;**19**(47):8986-8995

[50] Park J-M, Jeong M, Kim E-H, Han Y-M, Kwon SH, Hahm K-B. Omega-3 polyunsaturated fatty acids intake to regulate *Helicobacter pylori*-associated gastric diseases as nonantimicrobial dietary approach. BioMed Research International. 2015;**2015**:11

[51] Dyall SC. Methodological issues and inconsistencies in the field of omega-3 fatty acids research. Prostaglandins, Leukotrienes and Essential Fatty Acids. 2011;**85**(5):281-285

[52] Hadian Z. A review of nanoliposomal delivery system for stabilization of bioactive omega-3 fatty acids. Electronic Physician. 2016;**8**(1):1776-1785

[53] Thamphiwatana S, Gao W, Obonyo M, Zhang L. *In vivo* treatment of *Helicobacter pylori* infection with liposomal linolenic acid reduces colonization and ameliorates inflammation. Proceedings of the National Academy of Sciences of the United States of America. 2014;**111**(49):17600-17605

[54] Yashodhara BM, Umakanth S, Pappachan JM, Bhat SK, Kamath R, Choo BH. Omega-3 fatty acids: A comprehensive review of their role in health and disease. Postgraduate Medical Journal. 2009;**85**(1000):84-90

[55] Das UN. Essential fatty acids—A review. Current Pharmaceutical Biotechnology. 2006;**7**:467-482

[56] Manjari V, Das UN. Oxidant stress, anti-oxidants, nitric oxide and essential fatty acids in peptic ulcer disease. Prostaglandins, Leukotrienes and Essential Fatty Acids. 1998;**59**(6):401-406

[57] Obonyo M, Zhang L, Thamphiwatana S, Pornpattananangkul D, Fu V, Zhang L. Antibacterial activities of liposomal linolenic acids against antibiotic-resistant *Helicobacter pylori*. Molecular Pharmaceutics. 2012;**9**(9):2677-2685

[58] Jung SW, Thamphiwatana S, Zhang L, Obonyo M. Mechanism of antibacterial activity of liposomal linolenic acid against *Helicobacter pylori*. PLoS One. 2015;**10**(3):e0116519

[59] Li XX, Shi S, Rong L, Feng MQ, Zhong L. The impact of liposomal linolenic acid on gastrointestinal microbiota in mice. International Journal of Nanomedicine. 2018;**13**:1399-1409

[60] Nostro A, Cellini L, Di Bartolomeo S, Di Campli E, Grande R, Cannatelli MA, et al. Antibacterial effect of plant extracts against *Helicobacter pylori*. Phytotherapy Research. 2005;**19**:198-202

[61] Cowan MM. Plant products as antimicrobial agents. Clinical Microbiology Reviews. 1999;**12**(4):564-582

[62] Wu S, Chappell J. Metabolic engineering of natural products in plants; tools of the trade and challenges for the future. Current Opinion in Biotechnology. 2008;**19**(2):145-152

[63] Wang G, Tang W, Bidigare RR. Terpenoids as therapeutic drugs and pharmaceutical agents. In: Zhang L, Demain A, editors. Natural Products Drug Discovery and Therapeutic Medicine. Totowa, NJ: Humana Press; 2005. pp. 197-227

[64] Jäger S, Trojan H, Kopp T, Laszczyk MN, Scheffler A. Pentacyclic triterpene distribution in various plants—Rich sources for a new group of multi-potent plant extracts. Molecules (Basel, Switzerland). 2009;**14**:2016-2031

[65] Sheng H, Sun H. Synthesis, biology and clinical significance of pentacyclic triterpenes: A multi-target approach to prevention and treatment of metabolic and vascular diseases. Natural Product Reports. 2011;**28**(3):543

[66] Furtado NAJC, Pirson L, Edelberg H, Miranda LM, Loira-Pastoriza C, Preat V, et al. Pentacyclic triterpene bioavailability: An overview of *in vitro* and *in vivo* studies. Molecules. 2017;**22**(3):400

[67] Paraschos S, Magiatis P, Mitakou S, Petraki K, Kalliaropoulos A, Maragkoudakis P, et al. *In vitro* and *in vivo* activities of chios mastic gum extracts and constituents against *Helicobacter pylori*. Antimicrobial Agents and Chemotherapy. 2007;**51**(2):551-559

[68] Shin S-J, Park C-E, Baek N-I, Chung IS, Park C-H. Betulinic and oleanolic acids isolated from *Forsythia suspensa* Vahl inhibit urease activity of *Helicobacter pylori*. Biotechnology and Bioprocess Engineering. 2009;**14**(2):140-145

[69] Parreira P, Soares BIG, Freire CSR, Silvestre AJD, Reis CA, Martins MCL, et al. *Eucalyptus* spp. outer bark extracts inhibit *Helicobacter pylori* growth: *In vitro* studies. Industrial Crops and Products. 2017;**105**:207-214

[70] Jiang Q, Yang X, Du P, Zhang H, Zhang T. Dual strategies to improve oral bioavailability of oleanolic acid: Enhancing water-solubility, permeability and inhibiting cytochrome P450 isozymes. European Journal of Pharmaceutics and Biopharmaceutics. 2016;**99**:65-72

[71] Yang G, Yang T, Zhang W, Lu M, Ma X, Xiang G. *In vitro* and *in vivo* antitumor effects of folate-targeted ursolic acid stealth liposome. Journal of Agricultural and Food Chemistry. 2014;**62**(10):2207-2215

[72] Ge ZQ, Du XY, Huang XN, Qiao B. Enhanced oral bioavailability of ursolic acid nanoparticles via antisolvent precipitation with TPGS1000 as a stabilizer. Journal of Drug Delivery Science and Technology. 2015;**29**:210-217

[73] Fair RJ, Tor Y. Antibiotics and bacterial resistance in the 21st century. Perspectives in Medicinal Chemistry. 2014;**6**:25-64

[74] Miller LH, Su X. Artemisinin: Discovery from the Chinese herbal garden. Cell. 2011;**146**(6): 855-858